Exercises in Practical Astronomy using Photographs

with solutions

Exercises in Practical Astronomy using Photographs

with solutions

M T Brück

Formerly Senior Lecturer in Astronomy,
University of Edinburgh

Foreword by Owen Gingerich
Harvard-Smithsonian Center for Astrophysics, Cambridge, Mass.

Adam Hilger
Bristol, Philadelphia and New York

British Library Cataloguing in Publication Data

Brück, M. T. (Mary Teresa), *1925–*
Exercises in practical astronomy using photographs: with solutions.
1. Astronomy. Mathematical aspects
I. Title
520.151

ISBN 0-7503-0061-2

US Library of Congress Cataloging-in-Publication Data

Brück, M. T. (Mary T.)
Exercises in practical astronomy using photographs: with solutions/M. T. Brück.
p. cm.
Includes bibliographical references and index.
ISBN 0-7503-0061-2
1. Astronomy—Problems, exercises, etc. 2. Astronomy—Study and teaching. 3. Astronomical photometry. I. Title.
QB62.5.B78 1990
522′.076—dc20 90-41609
CIP

Consultant Editor: Professor A E Roy

Published under the Adam Hilger imprint by IOP Publishing Ltd
Techno House, Redcliffe Way, Bristol BS1 6NX, England
335 East 45th Street, New York, NY 10017-3483, USA

US Editorial Office: 1411 Walnut Street, Philadelphia, PA 19102

Typeset by BC Typesetting, Bristol BS15 5YS
Printed in Great Britain by The Bath Press, Avon

Contents

Foreword

Mary Brück's practical exercises in astronomy, based on the Edinburgh University Teaching Packages, have struck a responsive chord in me, both because they exemplify an approach to laboratory materials that I heartily endorse, and because they furnish a nostalgic reminder of my early days of teaching when such materials were not easy to find.

When I was a young graduate student in the 1950s, I was called to an assignment at a venerable teaching observatory in the Middle East. At the end of the nineteenth century the new 12-inch Brashear refractor at the American University of Beirut (which was then still the Syrian Protestant College) represented a substantial part of the institution's total investment. It was complete with the latest spectrograph, similar to the one described by James Keeler in an early volume of the fledgling *Astrophysical Journal*. But by 1955 the AUB had become a victim of neglect enhanced by World War II and by the absence of an active astronomer for many years. I dusted off the instruments, took the accumulating journals out of their wrappers, and began to discover a variety of intriguing pieces of teaching equipment dating back to earlier decades.

Among the teaching devices were split-lens assemblies for measuring the diameter of the Moon, charts for plotting the Moon and the planets, and small white 'Willson hemispheres' which together with crumbling spherical protractors could be used to solve graphically problems of spherical astronomy. I incorporated as much as I could fathom into a series of experiments for weekly lab sessions, and soon discovered that these efforts met with mixed success. Observational exercises required evening sessions and these were frequently thwarted by clouds. Even when the weather was good, the observations often proved difficult to carry out: plotting the course of the Moon with sufficient accuracy to determine where its path crossed the ecliptic, for example, was easier said than done because the glare of the Moon's full phase made it difficult to see appropriate reference stars. On the other hand, extensive plotting of pre-assembled data bored the students, and was more like a homework assignment than actual laboratory work. Among the exercises that worked well, however, was one that used a series of lunar photographs taken by John Duncan; here the students could ultimately discover the eccentricity of the Moon's orbit by measuring the varying size of the Moon's image.

After I returned to America, I discovered another of John Duncan's photographic exercises at Wellesley College, a series of globular cluster photographs that he had taken at Mount Wilson for finding the light-curve of variable stars. The exercise has not been widely distributed because it seemed to require the use of actual photographic prints. However, the progress of printing technology meant that much finer screens were available for half-tone reproductions, and it was indeed feasible to reproduce Duncan's plates with sufficient resolution so that the magnitudes of cepheid variables could be estimated from the printed page. This possibility led to a series of laboratory exercises in astronomy that for many years appeared from time to time on the pages of *Sky and Telescope*.

The use of photographic materials to take students closer to the original sources of astronomical data became a prime consideration when the secondary school Project Physics curriculum was developed at Harvard in the late 1960s. Fletcher Watson, the astronomer on the project's team of managers, strongly advocated such an approach, and he encouraged the use of actual photographs whenever possible. For example, he insisted that an exercise on triangulation of the place of Mars' orbit (in the style of Kepler) should begin with the measurement of the actual geocentric position of Mars from copies of photographs in the Harvard Observatory's plate stacks. We also wanted students to determine the eccentricity of the earth's orbit by measuring images of the Sun throughout the year. This exercise is similar to Duncan's lunar one, except that variation in size of the images is much less—so much less, in fact, that we discovered that the ellipticity of the solar images on our individual prints was greater than the annual effect being sought. We traced the difficulty to the fact that the ordinary photographic enlargement paper used in those days shrank much more in one direction as it was dried than at right angles. Ultimately we were forced to use the plastic paper that is commonplace today but which was quite unusual then.

This brief and spotty history of astronomy exercises simply documents the fact that improved technology has made possible increasingly better materials. The Edinburgh teaching resources that Dr Brück has made available in this volume represent yet another advance in our access to the kind of photographs astronomers actually use in their research. The images come from telescopes at the cutting edge of celestial investigations, many from the 1.2 meter UK Schmidt Telescope in New South Wales, Australia—in fact the series of exercises began with surplus transparencies of the very plates that the scientists were using for their trail-blazing studies. Here you will find photographs of such up-to-date phenomena as Halley's comet in its most recent apparition, and SN1987A, the great supernova in the Large Magellanic Cloud discovered in February 1987. Dr Brück's practical exercises include asteroids, star clusters and galaxies.

What will the future hold? Increasingly photographs are being replaced by digitized television-like scans of the remarkable objects found in the heavens. The Micro Observatory project now under development envisions an affordable remote controlled telescope that can send digitized images to the classroom, even in the daytime. Images obtained during clear weather can be stored for use on cloudy days, just as nighttime data can be stored for daytime use. With such equipment, students could not only repeat previous exercises, but could make real time observations of variable stars, of asteroids, or of sunspots.

While the Micro Observatory approach may duplicate the excitement of the research astronomer with his telescope, a view of the sky in squares of only half a degree on the side will not capture the glorious sweep of the skies, and the thrill of finding an esoteric nebula among the vast wilderness of stars. This is why I am so pleased that Mary Brück has recommended that her collection of exercises be used in conjunction with one of the modern star atlases. As much as any telescope, the wide-field Schmidt records the broad canvas of the heavens and, together with the atlas guides, these pictures not only provide an entry into practical astronomy, but they convey some of the majestic beauty of the starry firmament.

Owen Gingerich
Harvard-Smithsonian Center for Astrophysics

Preface

Practical work in astronomy at elementary and intermediate levels presents a difficulty for teachers. Unlike laboratory sciences, astronomy does not easily lend itself to bench experiments. Actual telescopic observations—even in climates where weather permits them—are usually limited to the Moon and planets. Meaningful observations of stars or galaxies, to match the instruction received in the classroom or from the textbook, are next to impossible.

Astronomical photographs supply an answer; they show what celestial objects actually look like, and are at the same time capable of being used to set problems relevant to the course work.

In these exercises, prints of first-class original photographs are reproduced. The photographs—except for those of the Sun—are negatives (black images on a clear background), as used by astronomers in their researches, on which objects can be identified and classified, their dimensions measured or their numbers counted. To use the photographs effectively requires only the most basic equipment. Though the methods of study are simple, they are in principle the same as those used by astronomers with their more elaborate resources.

It is assumed that students who use the book have a knowledge of astronomy at the level of a normal first year college course, or that they are currently attending such a course. A modest level of mathematics is required, specifically the use of logarithms and of simple trigonometrical functions. A knowledge of physics, beyond the general descriptive level of standard elementary astronomy textbooks, is not called for. Spectroscopy does not form a part of the programme of the exercises, except as a means of measuring motions.

Twelve topics have been chosen. Each one is briefly introduced, and enough additional information is given with each exercise to produce a satisfactory result. The topics are to a large extent independent in that it is not necessary to carry out the exercises in succession, though they have been arranged to follow the sequence of many elementary college courses.

The standard international (SI) system is used throughout for data and conversions which are given in the appendices.

M T Brück
July 1990

To the Teacher

It is intended that students should be able to carry out all or most of these exercises on their own. Possible methods of procedure are outlined for anyone who may find difficulty with one or other of the problems though they are by no means the only routes to a reasonable answer. Solutions are also provided at the end of each chapter and the results adopted by astronomical researchers are given and discussed where appropriate. Each chapter deals with one main problem, but in many cases the stages involved are divided into shorter exercises, so that the student who so wishes may keep a check on his or her progress. Other problems will suggest themselves to the teacher which may be tackled with the photographic material and the data.

The photographs are oriented so that east is anticlockwise from north; north is conventionally placed at the top, but this was not possible in some instances where the photographs had to be printed lengthwise on the page. Where it is essential to know the orientation, the north is marked. The angular scale of the photograph is also indicated, when needed.

The coordinates of interesting objects such as galaxies, when mentioned in the text, refer to the equinox of the year 2000. There are cases, however, (e.g. in identifying the position of Halley's comet) when the coordinates at another epoch have to be used. Where equatorial coordinates are required in an exercise, right ascension is given as an angle as well as in units of time. The objects are specified by their numbers in catalogues. NGC stands for 'New General Catalogue' and I for 'Index Catalogue', both originally produced by J L E Dreyer between 1888 and 1908. The current version of the combined catalogues is 'NGC 2000' (Cambridge University Press and Sky Publishing Corporation 1988). M stands for Messier's old catalogue (1784) of 110 conspicuous objects.

In addition to the simple equipment, listed on the next page, which each individual student ought to have, it is recommended that the class should possess:

(i) A globe for measuring angles on the celestial sphere. A celestial globe is useful, but not necessary. A school geographical globe serves just as well.

(ii) A good star atlas, for example Sky Atlas 2000.0 (W Tirion, Cambridge University Press and Sky Publishing Corporation 1981), Uranometria 2000.0 (W Tirion, B Rappaport and G Lovi, Richmond 1987), Norton's 2000.0 (ed. I Ridpath, Longman, Harlow and John Wiley, NY 1988) or Atlas Coeli 1950.0 (A Bečvář, Prague 1956). Though not essential for the exercises, an atlas on which regions in the sky may be looked up adds greatly to a student's enjoyment.

To the Student

The equipment you need is:

a millimetre ruler (which could be a strip of millimetre graph paper)
a compass
a protractor for measuring angles
a 10 × magnifier, if possible with a graticule divided into tenth millimetres
sheets of transparent overlay for tracing objects on the photographs
millimetre graph paper
a hand calculator which gives trigonometrical functions and base 10 logarithms
the use of a globe.

The preamble to each exercise gives you all the information you need. To begin with, you should ignore the 'Suggestions' and try to work out a method of finding an answer by yourself. If you succeed, you may check your results against the solutions at the end of each chapter. If, however, you are baffled, read the suggestions and try again.

One of the pitfalls is the conversion of data from one set of units to another, especially when dealing with data involving more than one dimension, such as velocity. Use SI units (metre, kilogram, second) for non-astronomical quantities. Never mix two types of units, or use powers such as kilometres or kiloparsecs in calculations.

Do not expect your results to tally exactly with the given solutions. Remember that astronomy is an observational science, and that different observers may get different results, depending on the accuracy of their individual measurement and on their interpretation of the material. Do what the astronomers do: make the most of your observations, assess the accuracy (or inaccuracy) of your result, and then stand by it!

Acknowledgments

I am very grateful to Professor M S Longair, Astronomer Royal for Scotland and Professor of Astronomy at the University of Edinburgh, an enthusiastic supporter of astronomy education, for encouraging me to put these exercises into book form. My warm thanks also go to my friends and colleagues at the Royal Observatory Edinburgh and the University of Edinburgh who shared their knowledge with me over many years of teaching in the Department of Astronomy.

The book would not have been possible without the beautiful photographs. I have been fortunate in being able to select many of them from the magnificent UK Schmidt Telescope Plate Library at the Royal Observatory Edinburgh. I am most grateful to the members of the UK Schmidt Telescope Unit for their co-operation especially Ms S B Tritton for her expert advice and guidance, and Mr M Read who selected the galaxies in the montage. I have been fortunate in the generous co-operation and interest of Mr Brian Hadley and his staff of the Royal Obvservatory Edinburgh Photolabs, who provided prints of the Schmidt Telescope photographs, many specially prepared for these exercises. I owe my thanks to Professor Ch. Fehrenbach for the spectrum of the Magellanic Cloud star and to Dr Ricard Casas i Rodriguez who kindly provided the beautiful photograph of the sunspot of March 1989. My warm thanks are also due to Mrs M Fretwell for her careful drawing of the diagrams.

1 The Sun

The Sun, our nearest star, is 1.5×10^{11} m from Earth. This distance, called the astronomical unit (AU), is one of the units used in astronomy, especially for distances inside the Solar System. The Sun has a radius of 7.0×10^{8} m. The bright disk which we see in the sky (plate 1.1) is called the *photosphere* or light sphere. A conspicuous feature of the photosphere is the decrease in its surface brightness from the centre to the edge or limb of the disk, a phenomenon known as limb darkening, which is a result of the gaseous nature of the photosphere whose temperature varies with depth. Dark sunspots of various sizes are another feature; they are temporary phenomena which reach a maximum in numbers approximately every 11 years.

Plate 1.1 The Sun showing limb darkening and a number of sunspots (© Royal Greenwich Observatory and Royal Observatory, Edinburgh).

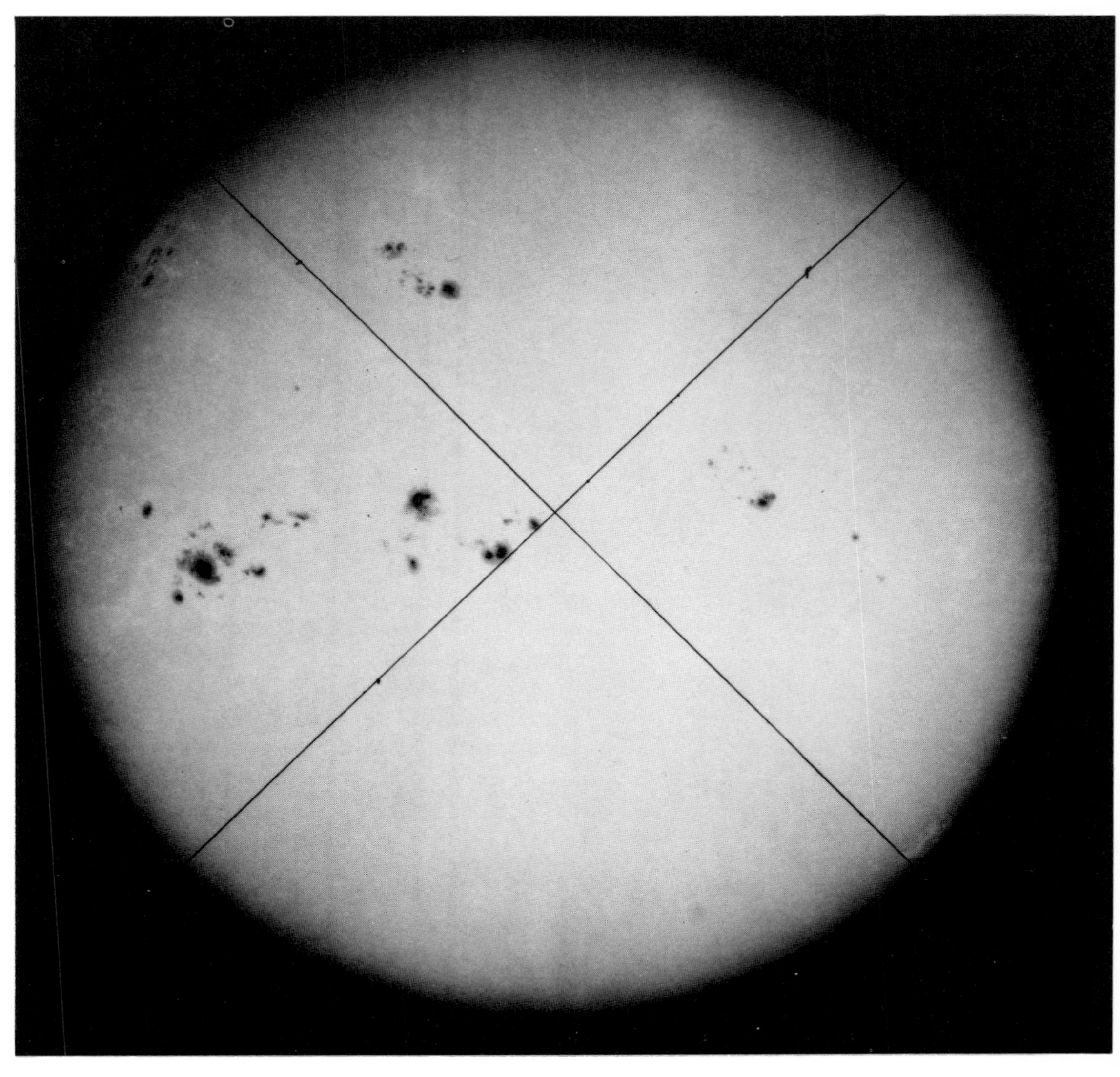

THE ROTATION OF THE SUN

The rotation of the Sun was first recognised from the movement of sunspots across its disk. The series of photographs (plate 1.2) taken over a period of time illustrates this clearly. If a particular spot persists for more than a complete rotation, the interval between two successive occasions when it is exactly on the central meridian (the diameter joining the Sun's poles on its projected image) is the period of rotation. However, it is possible to work out the period without following a spot through a whole cycle—indeed the lifetime of all but the largest spots is less than a cycle. The period observed in this way is the *synodic* period, that is, the period as seen from the Earth. Since the Earth is in motion around the Sun this period differs from the true or *sidereal* period which is that which would be observed from a stationary place in outer space.

The Sun's axis of rotation fixes its poles and its equator. Latitude on the Sun, as on the Earth, is measured from the equator. Longitude on the Sun is measured for convenience from the central meridian, though of course this meridian changes continuously as the Sun rotates. The plane of the Sun's equator is tilted at a small angle (7°) to the ecliptic, the plane of the Earth's orbit. As a result the Sun's axis of rotation is not normally in the plane of the sky. One of the poles is likely to be behind the visible disk at any moment. The Sun's axis of rotation is likewise not normally perpendicular to the ecliptic (figure 1.1).

The plane of the ecliptic is inclined at an angle of 23.4° to the plane of the Earth's equator; the angle which the axis of the ecliptic makes with the north–south direction in the sky varies therefore with the time of year. The actual angle between the axis of rotation of the Sun and the north–south direction in the sky (called the position angle of the Sun's pole) is a combination of the effect of the 23.4° inclination of the ecliptic and of the 7° inclination of the Sun's axis. The angle is calculated for every day in the year and published in advance in tables. It is easily observed from the apparent path of sunspots across the face of the Sun.

Sunspots are found in belts of solar latitude between 5° and 30° in both hemispheres. The path of a spot around the Sun's axis therefore follows a small circle of solar latitude. When the Sun's axis of rotation is exactly in the plane of the sky (in December and June, figure 1.1) the projection of the small circle is a straight line. At other times the apparent path is slightly curved, but for simplicity in calculation we assume that the shape of a small circle is not very different from a straight line.

The photographs (plate 1.2) are taken from the famous Greenwich daily photographic records of the Sun, begun in 1874 and continued for over a century. The particular set reproduced here belongs to the high sunspot maximum of 1948 and shows an unusually large number of spots and groups of spots. The orientation of the sky is indicated on all the photographs by lines produced by a pair of crossed threads in the focal plane of the instrument. The north (which is at the top) is found by bisecting the angle

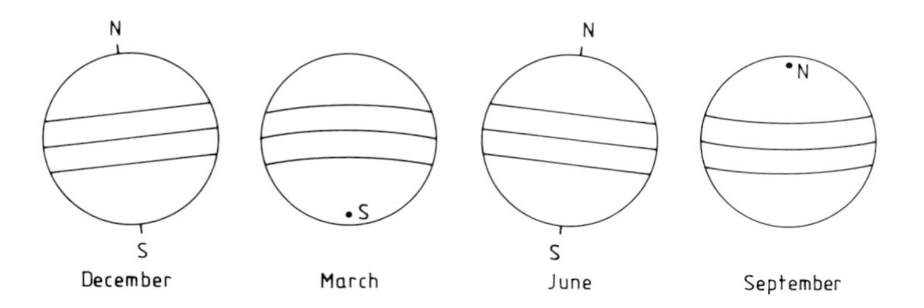

Figure 1.1 Aspects of the Sun in the course of the year.

between the cross-lines, though the intersection of the lines may not necessarily coincide with the centre of the Sun's image.

Exercise 1. Find the Sun's axis of rotation from the movement of sunspots across the disk.

Suggestions. Study the photographs and choose a spot whose motion can be followed over several days. The exercise may be repeated later using other spots, but it is advisable not to try to keep track of more than one spot at a time. Draw a circle on an overlay which has the same diameter as the Sun's image and mark on it a north–south diameter and also the centre of the circle. On each image bisect the angle between the crossed lines (by the use of a protractor) and draw a line parallel to this direction through the centre of the image. Place the overlay on the images successively, keeping the north–south line fixed, and trace the chosen spot. The tilt of the Sun's axis to the plane of the sky is so small (4° in the month of May) that the path of the spot is virtually a straight line. When these points are joined they mark the chord on which the particular spot moves. A line perpendicular to this through the centre of the circle marks the Sun's central meridian which should turn out to be the same whichever spot you use.

THE SUN'S SIDEREAL AND SYNODIC PERIODS

Figure 1.2(*a*) is a diagram showing the apparent path of a spot across the Sun's disk, which is a chord of the disk. The true path of the spot is a circle as shown in figure 1.2(*b*). The radius r of this circle is smaller than the Sun's radius R because the spot is not on the Sun's equator. If the spot is at longitude θ, the observer on Earth sees the spot at a projected distance $r \sin \theta$ (marked x on the diagram) from the central meridian. If r and $r \sin \theta$ are observed, the angle θ (its longitude) may be calculated for each date. The rate at which the longitude θ increases is the rate of the Sun's synodic rotation, and the time which the spot would take to travel 360° is the synodic period.

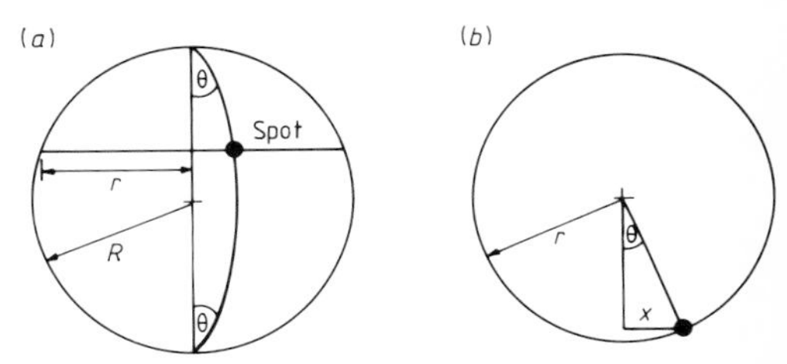

Figure 1.2 (*a*) Path of a sunspot on a chord of radius r. The sunspot is at longitude θ from the central meridian. (*b*) Cross-section through the Sun showing the circle on which the sunspot moves. The spot is seen at a projected distance x from the central meridian.

Figure 1.3 is a diagram in the plane of the ecliptic showing a cross-section of the Sun and also the orbit of the Earth. The directions of rotation of the Sun on its axis and of the Earth's motion around the Sun are shown. The Earth is at position 1 when a sunspot is seen on the central meridian at position A. After the Sun has performed a full rotation on its axis the spot has returned to position A but the Earth has moved on. The Sun has to rotate a little further until the spot appears when observed from the Earth to be once more on the central meridian. The Earth is now at position 2 and the sunspot at position B. The time it takes for the Sun to return to A is the sidereal period, which is the true period of rotation. The time it takes for the Sun to move from A back to A and on to B is the synodic period, in which it appears to perform a complete rotation as seen from the Earth.

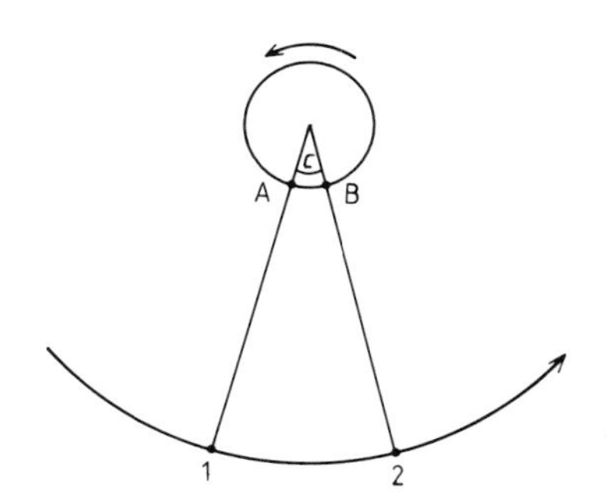

Figure 1.3 The plane of the ecliptic showing the rotating Sun and the Earth moving in its orbit, to illustrate the Sun's synodic and sidereal periods. A point on the Sun's equator is seen on the central meridian when the Earth is at position 1, and again when the Earth is at position 2. The interval between these two occasions is the synodic period during which the Sun has rotated through an angle $360 + c$ degrees and the Earth has moved an angle c degrees in its orbit.

If w is the rate at which the Sun rotates on its axis in degrees per day and W is the rate at which the Earth goes around the Sun in degrees per day, then a point on the Sun's surface gains $w - W$ degrees per day on the Earth. After one synodic period the gain is a full 360°. If the sidereal period of rotation of the Sun is P days and the period of the Earth around the Sun is Y days, then

$$w = 360/P \qquad \text{and} \qquad W = 360/Y \text{ degrees per day.}$$

If the synodic period is S days,

$$w - W = 360/S \text{ degrees per day.}$$

Putting the two equations together and dividing across by 360 gives the formula

$$1/S = 1/P - 1/Y$$

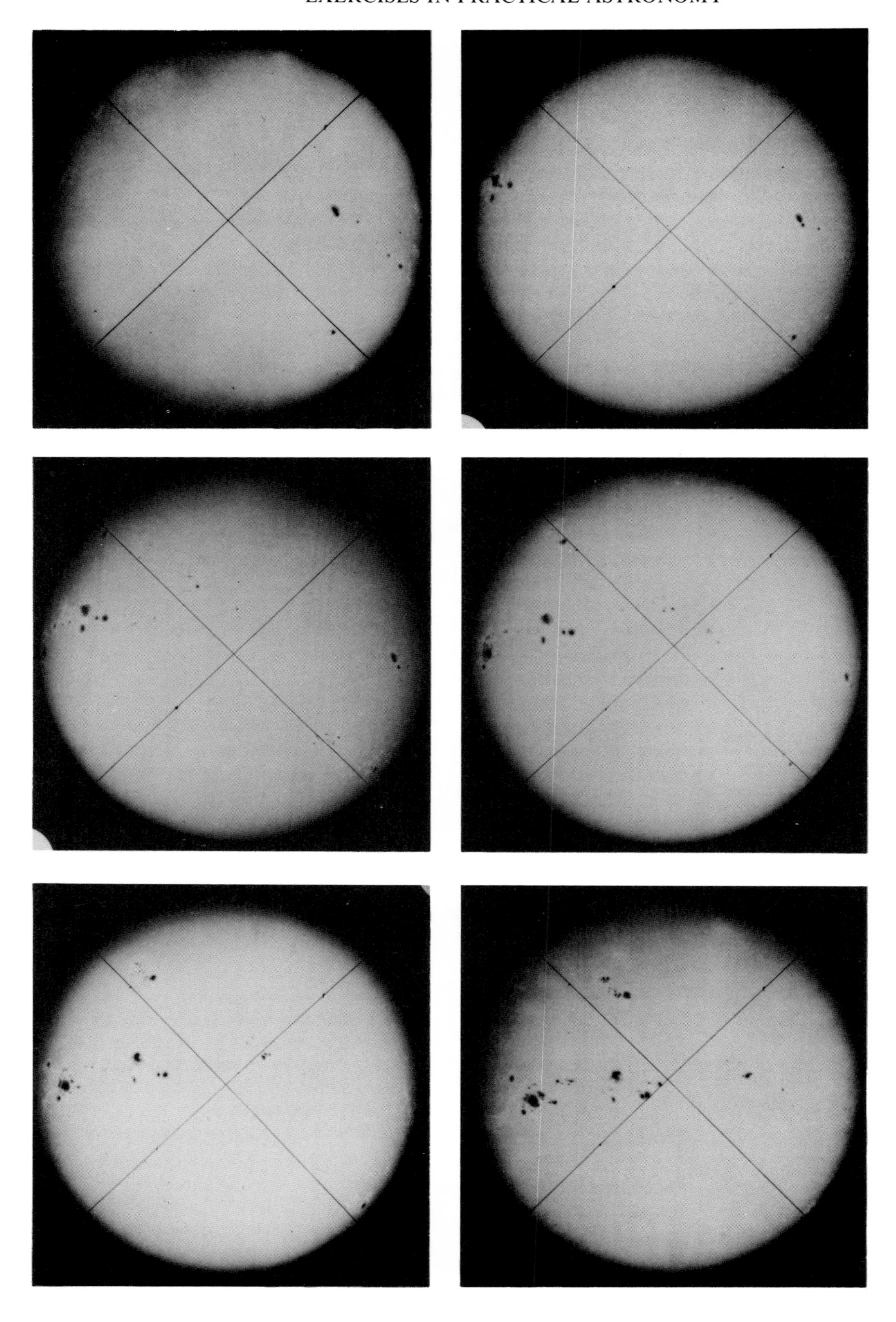

Plate 1.2 A series of daily photographs of the Sun, May 4–15, 1948 (© Royal Greenwich Observatory).

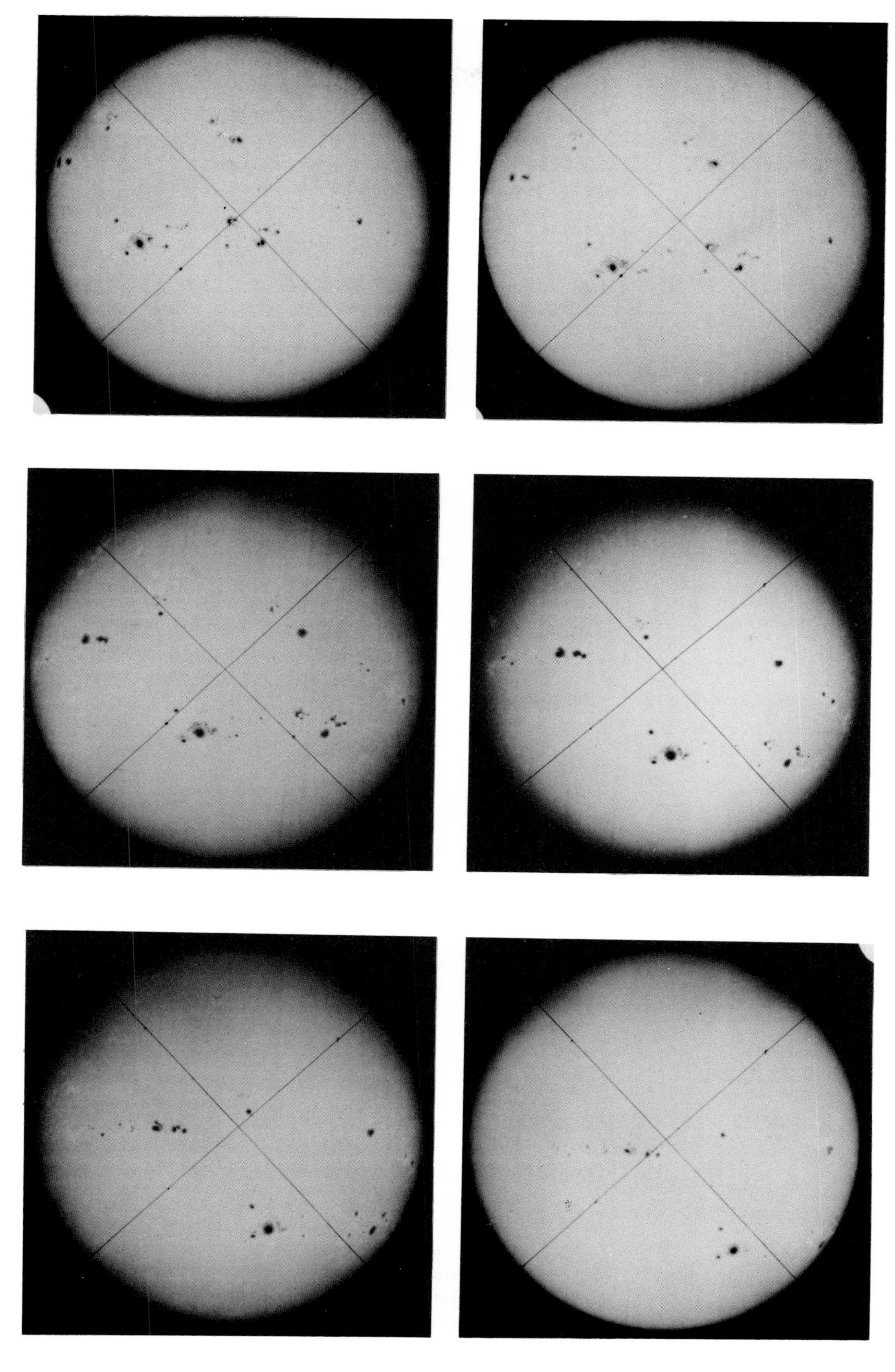

Y is the number of days in the year = 365.24.

Exercise 2. Find the Sun's synodic period of rotation from the motion of sunspots across the Sun's disk observed on the series of photographs. The photographs were taken close to noon, and the interval between successive photographs may be assumed, with sufficient accuracy, to be 1 day. Calculate the sidereal period from the synodic period.

Suggestions. The direction of the central meridian is the important basic line from which all positions are to be measured.

Beginning with the earliest date on which your chosen spot is visible, measure on each photograph (or on the overlay on which you have marked the successive positions of the sunspot) the perpendicular distance (x) of the spot from the central meridian along its chord. Measure also the diameter of the chord and divide by two. It does not matter which units are used (millimetres are the most convenient) as long as they are the same for both measurements, because only their ratio is needed. Divide the first by the second to get sin θ; then look up the value of θ. It is a good idea to enter your measurements in columns headed date, distance of spot, sin θ, θ. The angle θ is zero on the central meridian; make θ negative or positive according to whether the spot is to the east or the west of the central meridian.

Plot the last column against the first. Put the origin of your graph some days earlier than the first date. Leave room on the vertical ordinate for the angle θ to go from $-90°$ to $+90°$. Draw a straight line through the points and continue the line at both ends. Read off the dates which correspond to $+90°$ and $-90°$. The interval is the time it would take for the spot to travel 180 degrees or half the period.

The points may show a certain amount of scatter. One cause may be the changes in the spot itself over the time of observation. Another source of error may be your own measurements, and especially your value of r, because an error in this quantity will affect every point on your graph. Assuming that you have made your measurements as accurately as possible, there is a limit of precision fixed by the smallest distances which you can measure on the photographs. Note that the uncertainty in θ is greater when θ is large (when the spot is close to the edge or limb of the Sun) than when θ is small (when the spot is near the centre). Keep this in mind when drawing the straight line through the plotted points.

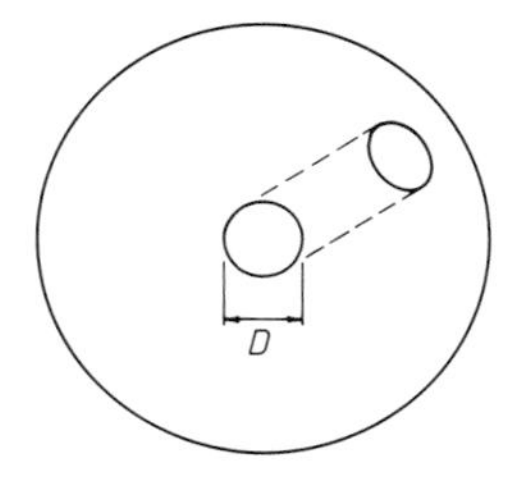

Figure 1.4 The effect of foreshortening. A circular feature of diameter D is seen in its true shape when at the centre of the disk, but appears as an ellipse when seen at any other position.

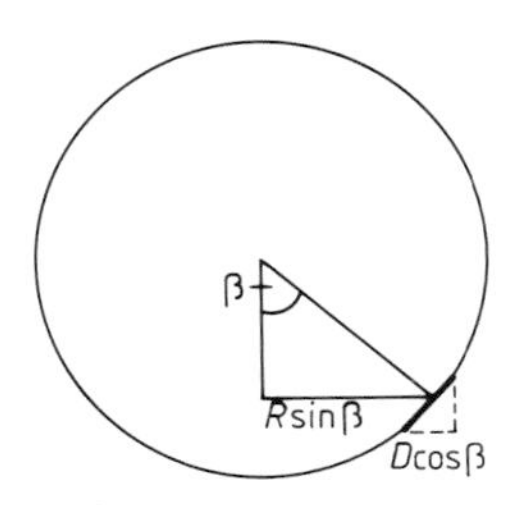

Figure 1.5 The foreshortened diameter of the feature in figure 1.4 is D cos β where β is its angular distance from the centre of the Sun. The other diameter remains D.

THE AREA OF A SUNSPOT

Sunspots change in size as they grow, reach maximum dimensions and then shrink and disappear. The study of a spot's development includes recording its area day by day. The apparent area requires correction for foreshortening because a spot appears smaller than its true size unless it is close to the centre of the Sun's image.

Figure 1.4 shows a round feature on the Sun at the centre of the disk which is seen in its true shape and size, and in another position where it appears foreshortened. However, the foreshortening occurs only in one dimension. The diameter at right angles to the Sun's radius is unshortened while that along the radius is shortened.

Figure 1.5 is a cross-section of the Sun through its centre in a plane in the line of sight. If the feature, of diameter D, is at an angle β from the line joining the centre of the sphere to the observer, its apparent diameter in this plane is D cos β. The diameter is unchanged in the perpendicular direction.

The angle β for a sunspot is found in a similar way to the longitude θ in Exercise 2. The apparent distance from the Sun's centre is R sin β where R is the radius of the Sun. The ratio of the distance from the centre to the radius is therefore sin β, from which cos β is calculated. One dimension of the spot is unforeshortened, the other is foreshortened by a factor cos β. The area, being the product of two dimensions, is foreshortened by the same factor as the foreshortened dimension, irrespective of the shape of the spot. The true area of a sunspot is therefore its apparent area divided by cos β. The areas of sunspots are often given for convenience in millionths of the Sun's visible hemisphere ($2\pi R^2$); this unit is equal to 3×10^{12} m^2.

Plate 1.3 is an example of a sunspot affected by foreshortening. The photograph, taken at the Observatorio del Teide, Tenerife, in March 1989, shows in detail a large and complex group of spots close to the limb of the Sun. The scale of the photograph is so large that the limb is almost a straight line. The central dark region of a sunspot is the *umbra* (Latin for shadow); the less dark region surrounding it is the *penumbra* or half-shadow. In the spot group observed on the photograph, a number of spots with their umbrae are enveloped in a large penumbra while a few small ones are separate from them.

Exercise 3. Calculate in millionths of the Sun's hemisphere (i) the area of one or more of the sunspots visible on plate 1.1 and (ii) the total area of the spot group including penumbra on plate 1.3. The longer dimension of plate 1.3 is 160 arcsec and the angular radius of the Sun on March 7, the date of the photograph, is 968 arcsec.

Plate 1.3 A part of the Sun photographed in green light showing the large sunspot of March 1989 when close to the limb. The umbra and penumbra are clearly seen. (© Instituto Astrofisica Canarias, La Laguna, Tenerife).

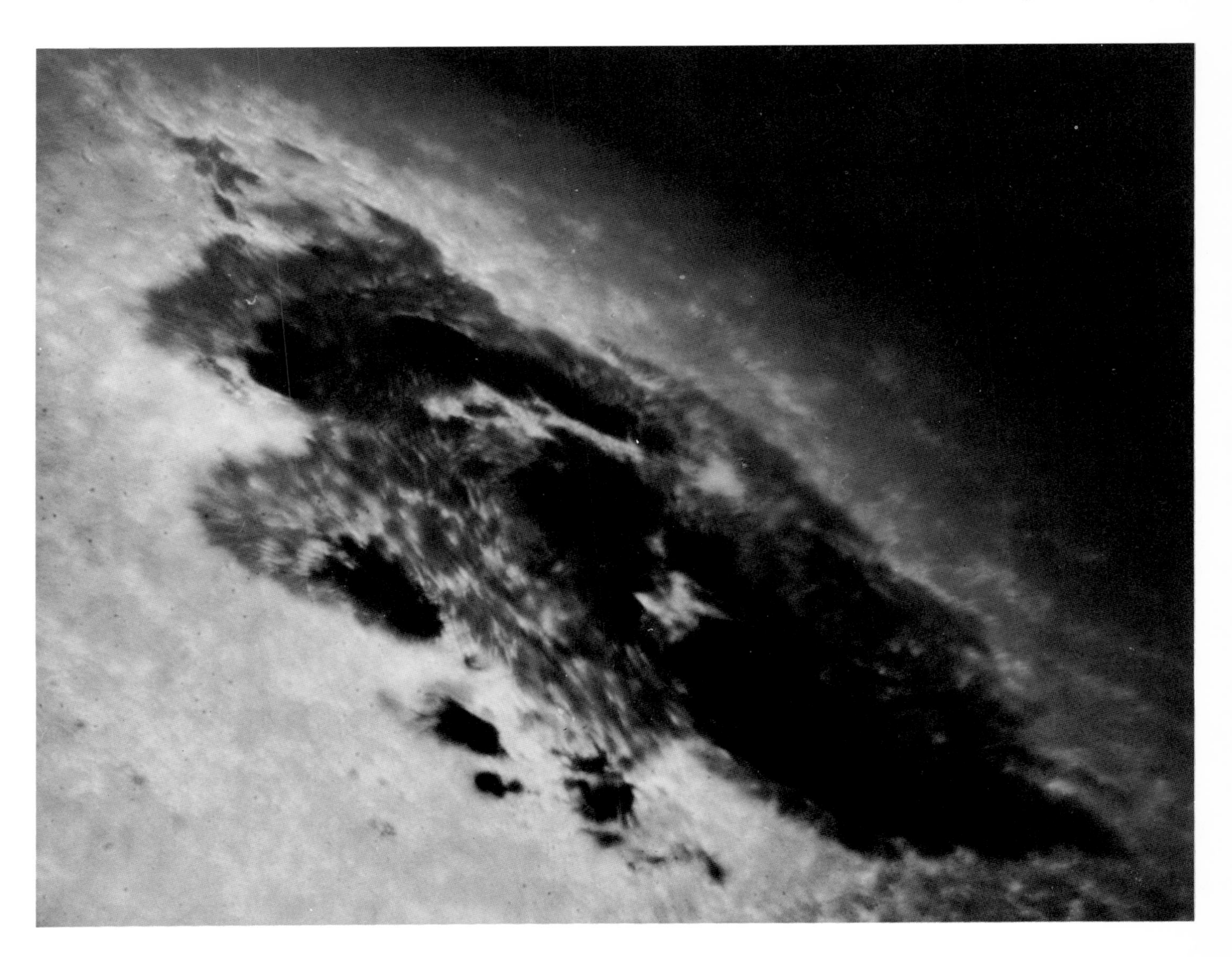

Suggestions. In case (i) measure the distance of the centre of the chosen spot from the centre of the Sun's disk as a fraction of the radius with a millimetre ruler, and calculate the foreshortening factor. Measure the area of the spot by tracing it on transparent squared graph paper (or by tracing it on an overlay which is then superimposed on a sheet of graph paper) and counting the number of squares. Calculate the area of the Sun's hemisphere in the same units. In case (ii) you will have to get the distance of the centre of the spot from the centre of the Sun's disk indirectly by first finding its distance from the limb of the Sun and subtracting this from the Sun's radius. The distance from the limb is the perpendicular distance, which is not difficult to find by the use of a square.

SOLUTIONS AND DISCUSSION

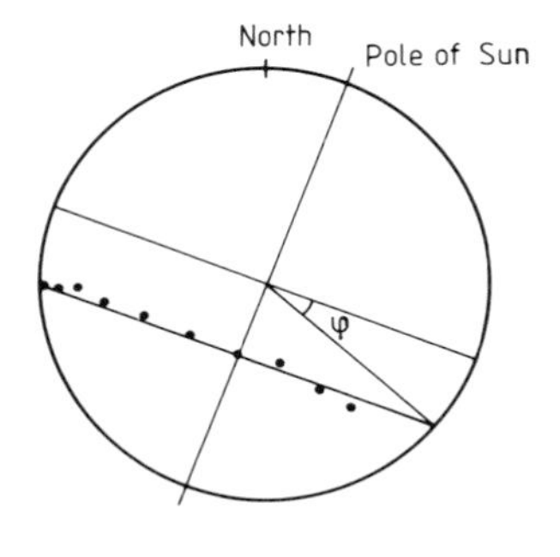

Figure 1.6 A sunspot at successive positions on the Sun's disk. The spot's latitude is φ.

1. Figure 1.6 is an example of the path of one of the sunspots across the disk. The spot is the one on the extreme east limb on the May 5 photograph, below the very large group. The same result should follow, whichever spot is chosen. The position angle of the axis of the Sun for May 4 is 23°; the angle varies from day to day, but the variation amounts to a decrease of only 2° between May 4 and 15. The fact that the positions of the spot do not fall perfectly on the chord is due in part to the difficulty of fixing the centre of the spot, which is not of constant size. Spots also drift slightly in latitude with time.

2. Table 1.1 gives a list of measurements made on the original Greenwich photographs on which the Sun's diameter is 140 mm and the semidiameter of the spot's path is 65 mm (the scale of the present photographs is smaller). Each value of x in the table is divided by 65 to give sin θ.

Table 1.1

Day	x	sin θ	θ (degrees)
2	−64	−0.985	−80
3	−60	−0.923	−67
4	−54	−0.831	−56
5	−43	−0.662	−41
6	−30	−0.462	−27
7	−16	−0.246	−14
8	−1	−0.015	−1
9	+13	+0.200	+12
10	+27	+0.415	+25
11	+38	+0.585	+36

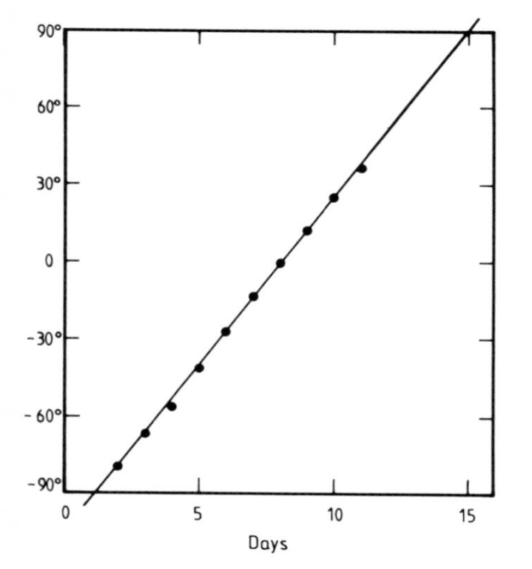

Figure 1.7 The spot's longitude plotted against time in days.

Figure 1.7 is a plot of the longitude θ of the spot against time in days. The first date in the series is Day 0; the particular spot used appears for the first time on Day 2 and is followed until Day 11. The straight line through the points crosses the −90° ordinate at Day 1.2 and the +90° ordinate at Day 15.0. The synodic period is the difference multiplied by 2, or 27.6 days. The answer should be reliable to within half a day. You may not get exactly the same value from your own observations, and you may well get different results for different spots. The reason for such variations is that the Sun does not rotate as a solid body: the rate of rotation is faster at the equator than at higher latitudes. The latitude of the spot (φ on figure 1.6) may be measured directly with a protractor from your drawing or calculated from the ratio r/R which is cos φ. In the example worked here φ is 22°S. The series of photographs has a conspicuous group of spots close to the equator. If you observe a spot in this group you will find that it moves more slowly across the Sun than those at higher latitudes. The group is a complex one, and you will need to take care to identify the same individual spot, as it changes its appearance from day to day.

The established periods of rotation are 26.9, 27.6 and 29.7 days at latitudes respectively 0°, ±20° and ±40°.

The formula which relates the Sun's sidereal and synodic periods can be rearranged in the form $1/P = 1/S + 1/Y$. Take the common denominator and invert:

$$P = S \times Y/(S + Y).$$

Using the value from the worked example of 27.6 for S and 365.2 for Y gives $P = 25.7$ days for the Sun's sidereal period.

On account of the fact that the Sun's rate of rotation varies with latitude it is customary for solar astronomers to use a sort of average rate when describing the Sun as a whole. That average synodic period is 27.3 days and the corresponding sidereal period is 25.4 days.

3. (i) As an example, the large spot in the left hemisphere lies 0.70 of the distance between the centre and the edge; therefore sin $\beta=0.70$, cos $\beta=0.71$ and the foreshortening correction is $1/0.71=1.4$. The area of the spot (on an image of radius 70 mm) was 11 mm^2 or 15.4 mm^2 corrected. The area of the Sun's hemisphere in the same units is 3.1×10^4 mm^2; the ratio of the areas is 5×10^{-4}, so the area of the spot is 500 millionths. (ii) The distance of the centre of the spot group from the edge of the Sun, converted to angular measure, is 46 arcsec; therefore its distance from the centre of the Sun (radius 968 arcsec) is 922 arcsec, a fraction 922/968 or 0.952 of the radius. Therefore sin $\beta=0.952$, cos $\beta=0.306$ and the foreshortening correction is 3.27. The area of the spot group (measured in square centimetres on squared graph paper and then converted to square arcseconds) is 5115 or 1.67×10^4 corrected for foreshortening. The ratio of this to the area of the Sun's hemisphere in the same units is 2.84×10^{-3} or 2840 millionths. As the scale of this photograph is much larger than that measured in (i), a more accurate value of the area is attained. There is some uncertainty involved in deciding what point to adopt as the centre of the group, but not a serious one, as you can discover by trying slightly different centres and seeing how the foreshortening correction factor is affected. This is an unusually large sunspot.

You may wish to follow the evolution of a sunspot by repeating this exercise on one particular spot over many days on the series of photographs. However, the scale is too small to give more than a rough result.

2 Minor Planets or Asteroids

Five planets visible to the naked eye—Mercury, Venus, Mars, Jupiter and Saturn—have been known from the dawn of history. In 1781 a new planet, Uranus, beyond the orbit of Saturn, was discovered by William Herschel. Two further planets beyond Uranus were later found, Neptune in 1846 and Pluto in 1930.

As well as these, there exist thousands of small planets which occupy orbits between those of Mars and Jupiter (figure 2.1). The first one was discovered on January 1, 1801 in Palermo by Giuseppe Piazzi and named Ceres after the tutelary goddess of Sicily. Others soon followed. The early ones were individually given names, mostly female, from the world of classical mythology such as Pallas, Vesta, Juno and Iris. Hundreds were revealed through visual searches and were listed by number in order of discovery, beginning with Ceres as number one. These small planets are called asteroids or minor planets. The word asteroid, which means a small star, is misleading, though widely used; it is preferable though perhaps clumsy to use the name minor planet. With the use of photography more than 10 000 minor planets have been discovered and are still being discovered. Their movement is so rapid that on a long exposure photograph their images appear as trails; indeed, this is how new ones are recognised. The photograph (plate 2.1) includes examples of such trails which will be used in this chapter as a demonstration of planetary motion.

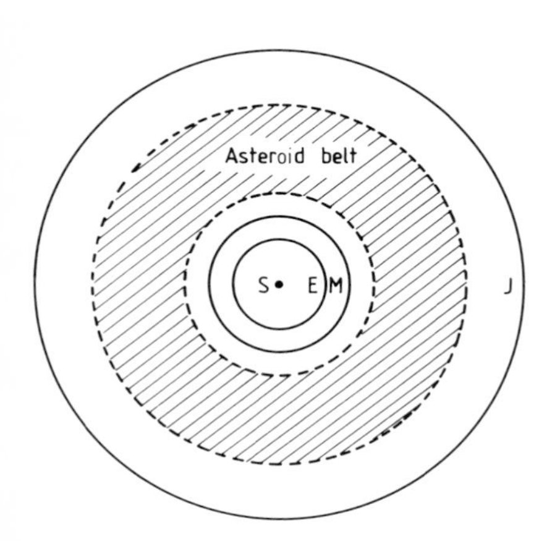

Figure 2.1 The orbits of the planets from Earth to Jupiter. The orbits of the minor planets or asteroids fall between those of Mars and Jupiter.

Plate 2.1 covers a small field in the plane of the ecliptic. The ecliptic is defined by the Sun's apparent annual path through the sky. As the Earth goes round the Sun, the Sun appears to make one complete revolution of the sky with respect to the stars. Its path is in the great circle of the ecliptic which is inclined at an angle of 23.4° to the celestial equator. At the points where it crosses the equator the ecliptic makes an angle of $+23.4°$ or $-23.4°$ to the celestial equator. Half-way between these points the angle is 0° because the ecliptic is parallel to the equator. At other positions the angle is between these limits.

The orbits of planets are governed by Kepler's laws. The first law states that planets move in ellipses with the Sun at one focus (see figure 3.1 for a diagram of an ellipse). The second law describes the rate of motion of the planet in different places on the ellipse; when the orbit is a circle that motion is uniform. The third law states that the square of the period (P) of a planet is proportional to the cube of the semimajor axis (half the longest axis of an ellipse, a) or, in the case of a circular orbit, the radius of the circle. The formula is

$$P^2 = ka^3.$$

Kepler's laws are explained by Newton's law of gravitation. The law of gravitation states that the force between two massive bodies is proportional to the product of the masses and inversely proportional to the square of the distance between them.

The law is written (for two masses M and m separated by a distance r):

$$F = GMm/r^2$$

Plate 2.1 Tracks of minor planets on a photograph exposed for 70 minutes (© Royal Observatory, Edinburgh).

where G is a constant called the constant of gravitation.

Starting from Newton's law of gravitation, Kepler's third law is derived in the form

$$P^2 = 4\pi^2 a^3 / G(M+m)$$

where P is the period in seconds, a is the semimajor axis in metres and M and m are the masses in kilograms of the Sun and the planet respectively.

If the units are changed, with the period given in years (the period of the Earth around the Sun), the distance in astronomical units (the distance of the Earth from the Sun) and the mass in units of the Sun's mass, the third law becomes

$$P^2 = a^3/(M+m).$$

In the solar system itself, $M=1$, and m, the mass of the planet, is so small compared to the mass of the Sun that it can generally be left out of the formula. Putting 1 for the mass leaves the third law in the form in which it was enunciated by its discoverer, Kepler:

$$P^2 = a^3.$$

The majority of minor planets have orbits which lie fairly close to the plane of the ecliptic. Their paths extend over a belt between roughly 2 and 4 AU from the Sun. Besides the 'belt asteroids' there are groups in quite different orbits, for example the 'Apollo asteroids' named after the first of their type, which move in highly eccentric orbits that cross the orbits of Mars and Earth.

The study of the orbits and numbers of minor planets is important for the understanding of the history of the solar system. If an orbit is a perfect circle, two observations of position and the time interval between the observations are sufficient to specify the orbit completely. The orbits of minor planets in the belt are not very eccentric, and as a first approximation may be assumed to be circular.

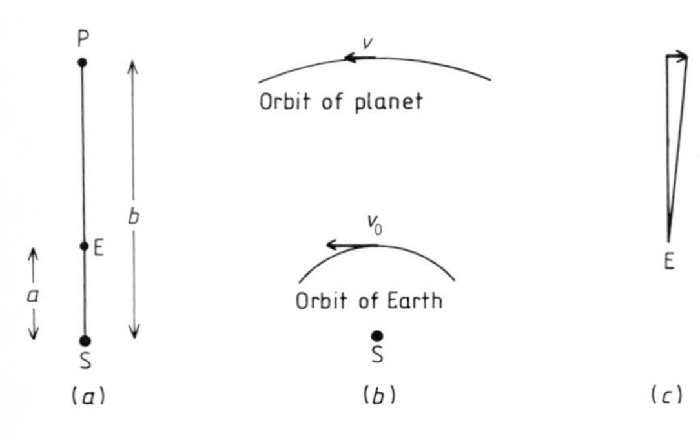

Figure 2.2 (*a*) The positions of the Sun (S), the Earth (E) and a planet (P) when the planet is in opposition. (*b*) Motions and directions of motion of the Earth and of the planet in their orbits at the time of opposition. (*c*) Apparent motion of the planet as seen from the Earth. The planet appears to move backwards (retrograde) because it is being overtaken by the Earth.

When a planet is 180° from the Sun in the sky, it is said to be *in opposition*, and the geometry of the configuration of Sun, Earth and planet is a very simple one (figure 2.2(*a*)). Figure 2.2(*b*) shows the motions of the Earth and the minor planet when in opposition. Their distances from the Sun are respectively a and b metres. The vectors show the velocities of the two bodies. The minor planet travels more slowly than the Earth, and to an observer on Earth it therefore appears to be moving backwards, or *retrograde* (figure 2.2(*c*)). The position of the minor planet at the beginning of the exposure is therefore the east or left side of the trail, and the velocity of the planet relative to the Earth is the difference between their two velocities.

The formula for the velocity of a planet in a circular orbit of radius a is

$$V = (GM/a)^{1/2}$$

which means that the velocity is inversely proportional to the square root of the distance from the Sun. (The formula is Kepler's third law expressed in another way, as is easily proved by substituting V for $2\pi a/P$ in the third law formula given earlier.)

If the velocities of the Earth and the asteroid are in m s^{-1}, the angular velocity W is (figure 2.2(*c*))

$$(V_0 - V)/(b-a) \text{ rad s}^{-1}.$$

Since the velocities are inversely proportional to the square root of the distance we have

$$V = V_0(a/b)^{1/2}.$$

Substituting this in the expression for angular velocity gives

$$W = V_0[1-(a/b)^{1/2}]/(b-a).$$

This formula can be simplified by a little algebra. For $(b-a)$ put $a(b/a-1)$. Factorise $(b/a-1)$ as the difference of two squares: $[(b/a)^{1/2}+1]\times[(b/a)^{1/2}-1]$. The value of V_0, the Earth's velocity, is $2\pi a/P$ where P is the length of the year in seconds, so $V_0/a=2\pi/P$.

The result is

$$W=(2\pi/P)/[(b/a)+(b/a)^{1/2}] \text{ rad s}^{-1}.$$

Expressed in words, the formula means that the length of a trail divided by the exposure time can provide a value for b/a, which is the distance of the planetary orbit from the Sun in AU.

Table 2.1 gives details of plate 2.1.

Table 2.1

Date of photograph	July 20, 1977
Coordinates of field	19h 57m (299.2°), −20.0°
Coordinates of Sun	7h 55m (118.8°), 20.8°
Length of exposure	70 minutes

Exercise 1. Make use of the information in table 2.1 to identify the belt asteroids on the photograph. The photograph is lined up as shown in the north–south direction to an accuracy of better than a degree.

Suggestions. It is obvious from the coordinates of the Sun and the field that the field is in opposition, being almost exactly diametrically opposite to the Sun in the sky. The belt asteroids move in the plane of the ecliptic. The problem is to find the direction of the ecliptic on the photograph. The east–west direction on the photograph is parallel to the equator. Consult a star atlas on which the ecliptic is shown to locate the position of the field from its coordinates. You will find that the field is on or very near the ecliptic. Measure with a protractor the angle which the ecliptic makes with the east–west direction on the atlas (you may find it easier to measure against the north–south direction to avoid the curved circles of declination).

Locate the trails on the photograph; draw a fine line in pencil on the photograph or on a transparent overlay along each trail and continue the line. Mark also on the overlay the east–west direction. An overlay of transparent graph paper is very useful for lining up purposes. Measure the angle between the trail and the horizontal. A trail which is parallel to the ecliptic is inclined at the same angle which you have measured from the atlas, but you should include any trails within 2–3° of this value as representing objects which move in or near the plane of the ecliptic.

Exercise 2. From the length of the trails and the information already given, calculate the distances of the belt asteroids from the Sun in AU.

Suggestions. The trails, being short, must be measured with the greatest available precision in order to get a reasonably accurate result. Measure them with a fine graticule or, failing that, with a millimetre ruler estimating to the nearest quarter or fifth of a millimetre by using a hand magnifier. From the given scale of the photograph convert the trail length to angular measure in arcseconds. Divide this by the exposure time in seconds to get the angular rate of motion of the asteroid in arcseconds per second.

Now recall the formula derived earlier for the asteroid's angular velocity:

$$W=(2\pi/P)/[(b/a)+(b/a)^{1/2}] \text{ rad s}^{-1}.$$

By multiplying by the number of arcseconds in a radian this is converted to arcseconds per second. P in the formula is the length of the year in seconds. (These numerical quantities are given in Appendix 1.)

If the angular velocity in the formula is put equal to the angular velocity which you have measured, you get an equation in b/a. This can be solved by algebra, but it is quicker to solve it by trial with a calculator. We know that the minor planet is beyond Mars, so try first whole numbers like 2, 3 or 4 to get the nearest whole number, then intermediate values until you get the correct solution.

SOLUTIONS AND DISCUSSION

1. The angle between the ecliptic and the east–west line is 11.5°. There are three asteroids which may be regarded as belt asteroids. A fourth trail is at quite a steep angle to the ecliptic. This is an object moving in an orbit of high inclination which is very likely also to be eccentric. Without further observations there is little to be learned from such a trail, other than to identify it as a non-belt member.

2. The trail lengths (measured by one observer) are 0.80, 0.90 and 0.67 arcmin. When the values of π, P and the number of arcseconds in a radian are substituted in the formula for W we get

$$W[(b/a)+(b/a)^{1/2}]=4.10\times 10^{-2}\ \text{arcsec s}^{-1}.$$

The first trail (0.80 arcmin in 70 min) is equivalent to 0.0114 arcsec s^{-1}. Putting this value of W in the formula:

$$[(b/a)+(b/a)^{1/2}]=3.60.$$

The value of b/a which (by trial and error) best fits this formula is 2.1. The asteroid is therefore 2.1 AU from the Sun. The distances of the other two, found in the same way, are 1.8 AU and 2.7 AU. It is evident from the actual process of measuring the lengths of the trails that there is uncertainty in the measurements, depending on the fineness of the scale or graticule used. If, for example, the uncertainty is 5% of the trail length in the case of the worked example, the value of $[(b/a)+(b/a)^{1/2}]$ could be anything from 3.42 to 3.78. If you repeat the calculation with these two figures, you will find values of 2.0 and 2.3 for b/a. For this reason it would be meaningless to give the result to more than two significant figures. In view of these uncertainties of measurement, do not be surprised if your answers differ from those worked out here.

On account of the thickness of the photographic images there is also a bias towards getting measurements for the trails which are slightly too large, thus making the distances of the asteroids slightly too small.

You may find it interesting to make a rough estimate of the total number of asteroids potentially visible on UK Schmidt Telescope photographs assuming that the small ecliptic field examined here is typical.

3 Halley's Comet

Comets are the most fascinating and attractive objects in the solar system. Though rarely bright enough to produce a spectacle to the naked eye, comets are quite frequent. About 10 are recorded each year, some new, and some expected from previous visitations. The comets which return regularly are called periodic comets. The new ones are strictly speaking also periodic but of such long periods that their earlier appearances have not been recorded; furthermore, comets may be disturbed by the gravitational influence of the planets during their long absence in more distant parts of the solar system, leaving their orbits unrecognised on their return.

The orbit of a comet is an ellipse which obeys Kepler's laws (§2). The dimensions and shape of an ellipse are described by the semimajor axis a and the eccentricity e. Another useful quantity is the closest distance to the Sun called the perihelion distance. Figure 3.1 shows these quantities. The perihelion distance, denoted by the letter q, is $a(1-e)$.

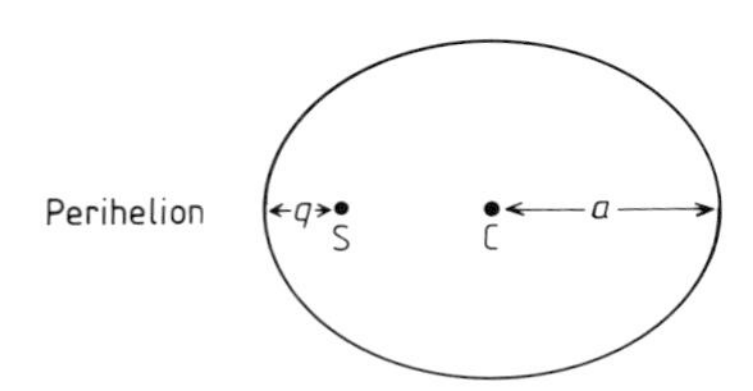

Figure 3.1 Diagram of an ellipse, the type of path followed by planets and comets around the Sun. C is the centre and the Sun (S) is at the focus. Perihelion is the point in the orbit nearest to the Sun, a is the semimajor axis and q is the perihelion distance.

Newton explained cometary orbits in terms of highly eccentric ellipses. Halley, on calculating the path of the magnificent comet of 1682, concluded that the great comets recorded in 1531 and 1607 (intervals of 76 years) were the same object, and foretold its return in the year 1758. The comet, named after him, returned as he had anticipated. Historical references to Halley's comet go back more than 2000 years. Its appearance in 1986 was its 30th on record.

Many comets have quite short periods and move in direct orbits, that is, in the same sense (eastwards) as the planets. Halley's comet is an example of a long period comet. Its period is 76 years, perihelion distance 0.6 AU, semimajor axis 18 AU and eccentricity 0.97. Its orbit is retrograde, that is, the comet moves in the contrary sense to the planets. The plane of the orbit is inclined at an angle of 18° to the ecliptic. Figure 3.2(*a*) shows the whole orbit projected onto the ecliptic, and figure 3.2(*b*) the part of the orbit near the Sun.

When the orbit of a comet is known it is a straightforward matter to calculate its position in space at any moment. Predicted positions are published in advance which give the comet's distance from the Sun, its distance from the Earth, its coordinates as seen from the Earth and other details.

A comet consists of a nucleus only a few kilometres across (not visible on the photographs) surrounded by a coma (a Greek word that means 'hair') or head. The tail is formed when the comet draws near to the Sun. When distant from the Sun the comet is a tenuous structure of dust and frozen gases, aptly described as a 'dirty snowball'. Its mass is very low, about 10^{-10} that of the Earth. On coming into the vicinity of the Sun the ices evaporate and the gases and dust are released to form the coma and the tail or tails.

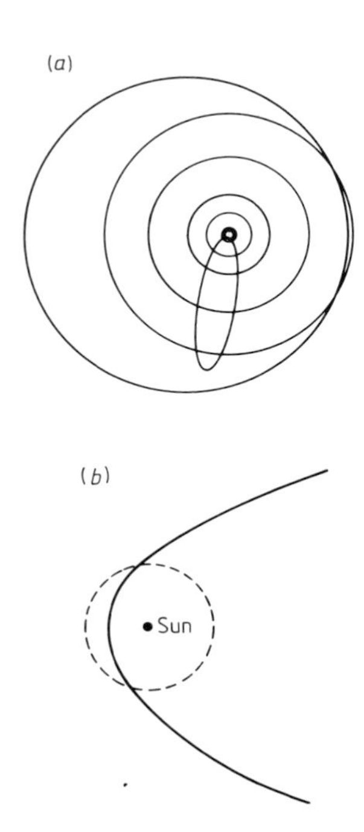

Figure 3.2 (*a*) The path of Halley's comet and the orbits of the planets from Mars to Pluto. (*b*) The part of the path of Halley's comet near the Sun and the Earth's orbit (dotted circle).

THE TAILS OF COMETS

A striking feature of comets' tails is that they always point away from the Sun. There are two types of tail, which are distinguished both by their appearance and by their spectra. Type I is a straight tail which points directly away from the Sun. Type II is a broad tail, or a number of tails, spread out like a fan. The spectrum of the broad tail is the spectrum of sunlight scattered by dust particles. The spectrum of the straight tail shows that it is composed of ions (electrically charged particles) of various molecules and radicals. This tail is not smooth but shows knots, waves and twists.

The dust tail is caused by pressure of the Sun's radiation which pushes the small dust grains away with a force which is stronger than the Sun's gravitational attraction. Light consists of photons of energy, and as such is capable of exercising pressure on anything which it strikes. The radiation pressure exceeds the gravitational attraction if the grains are very small—about the size of the wavelength of light. All other things being equal, small grains accelerate away faster than larger ones. Since the motion is an acceleration, each grain travels on a curved path with respect to the head of the moving comet. A cloud of grains of various sizes, having left the comet, spreads out into a fan in which the small grains travel farthest and straightest in any interval of time. The shapes of dust tails are sometimes more complex than this, however.

The mechanism which drives the ion tail is the solar wind. Indeed the solar wind was postulated to account for the ion tail many years before it was directly observed from space. The solar wind is a stream of plasma (ions and electrons) emanating from the Sun. The ions are mainly protons (nuclei of the hydrogen atom) because hydrogen is the most abundant element in the Sun. Instruments in space record 5×10^6 protons per cubic metre travelling outwards at a speed of about 5×10^5 m s^{-1} when recorded from space in the vicinity of the Earth. In a simple picture the ions in the comet, blown by the solar wind, may be imagined as being drawn out, like smoke from a chimney on a windy day, to form the straight tail.

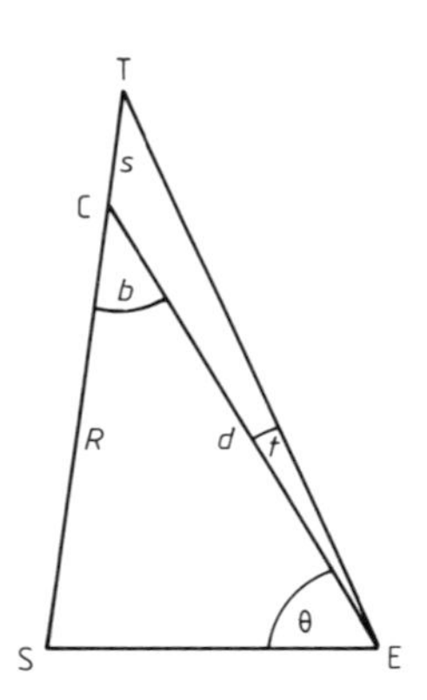

Figure 3.3 Configuration of Sun (S), Earth (E) and comet (C) with its tail extending to T.

Figure 3.3 shows the configuration of the Sun, Earth and comet in space. The extremity of the tail is on the continuation of the line joining the Sun to the comet. In the large triangle ESC the distance R (Sun to comet) is known from knowledge of the comet's orbit; the distance SE (Sun to Earth) is 1 AU. If the angles θ (between Sun and comet) and t (the apparent angular extent of the comet's tail) are observed, the length of the comet's tail may be found in astronomical units.

Plate 3.1 is a photograph of Halley's comet taken on March 10, 1986. The tail was so long that it required two photographs taken with the UK Schmidt Telescope to accommodate it. The composite photograph shows the junction between the two photographs which are inclined at a small angle to each other because of the curvature of the sky. Each photograph covers 6.5° × 6.5° of sky and the vertical sides point towards the pole.

Table 3.1 gives the positions in the sky of Halley's comet and the Sun when the photograph was taken.

Table 3.1

Date	March 10, 1986
Coordinates of comet	20h 11m (303°), −20°
Coordinates of Sun	23h 23m (351°), −4°

Exercise 1. Demonstrate, given the coordinates of the Sun and the comet in table 3.1, that the comet's ion tail points away from the Sun. The distance of the comet from the Sun on that date, computed in advance from the known particulars of its orbit, was 0.85 AU; calculate the length of the comet's tail in the same units from its angular extent on plate 3.1.

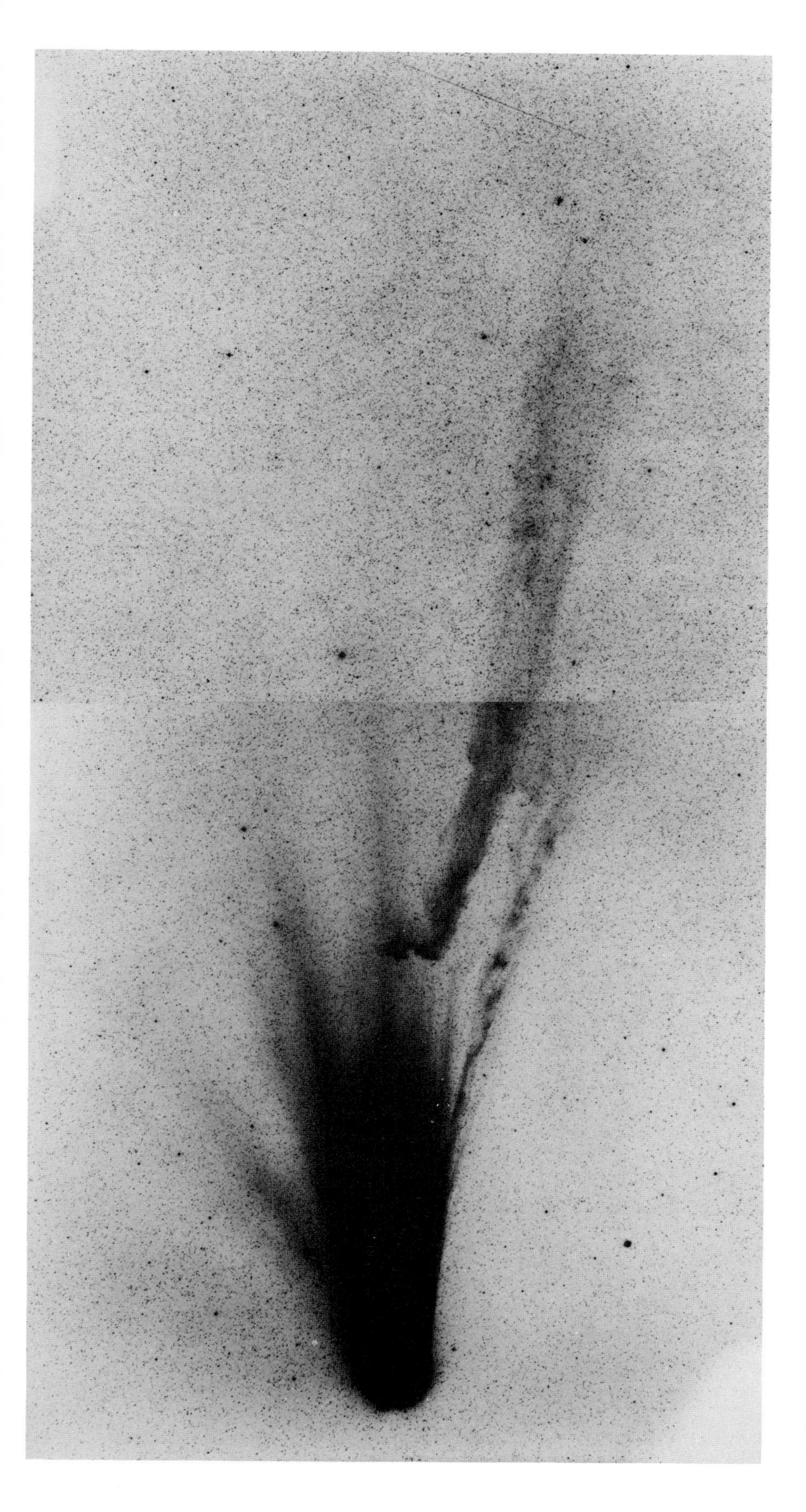

Plate 3.1 Halley's comet on March 10, 1986 showing the full extent of the tail (© Royal Observatory, Edinburgh).

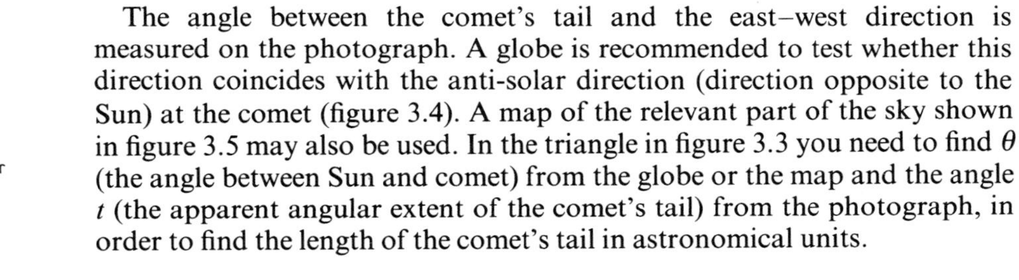
The angle between the comet's tail and the east–west direction is measured on the photograph. A globe is recommended to test whether this direction coincides with the anti-solar direction (direction opposite to the Sun) at the comet (figure 3.4). A map of the relevant part of the sky shown in figure 3.5 may also be used. In the triangle in figure 3.3 you need to find θ (the angle between Sun and comet) from the globe or the map and the angle t (the apparent angular extent of the comet's tail) from the photograph, in order to find the length of the comet's tail in astronomical units.

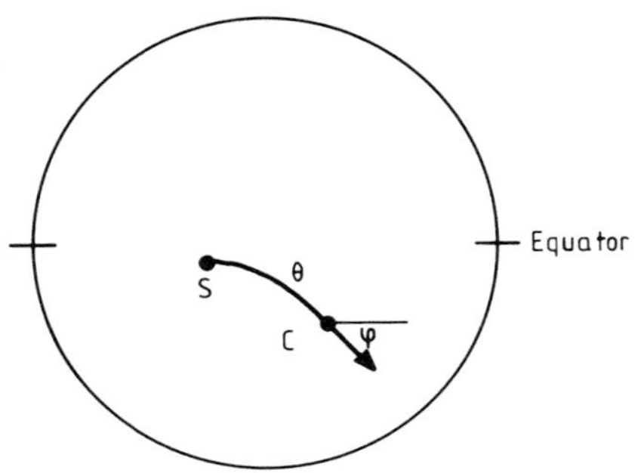

Figure 3.4 How to find the angular distance θ between the Sun and the comet from a globe. The arrow is the antisolar direction which is at an angle φ to a parallel of declination.

Suggestions. It is not difficult to measure the direction of the tail with a protractor by drawing an east–west line on the photograph through the head of the comet. Mark the positions of the Sun and the comet on the globe. Join the points by stretching a piece of string between them. Lay a protractor on the globe at the position of the comet and measure the angle (φ) between the direction of the string and the parallels of declination. Measure also the distance between the points on the string, or use a measuring tape (a long thin strip of millimetre graph paper makes an excellent and accurate measuring tape), and convert this length to an angle (θ) by comparing it with the circumference of the globe which corresponds to 360°. Alternatively, plot the positions of the comet and the Sun on the map (figure 3.5) and measure their separation.

The angle φ should be the same as the one which you have measured on the photograph. Halley's comet has a complex tail structure. When you have found the position angle φ from the globe, compare it with the directions of the various tails on the photograph to find which one most closely follows the anti-solar direction.

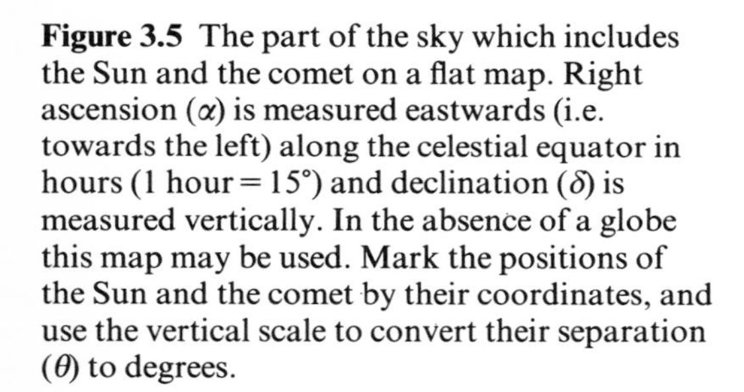

The next step is to find the length of the comet's tail in degrees. Measure this on the photograph in millimetres and convert it to degrees using the scale of the photograph provided by the outline of the individual square photographs, of side 6.5° exactly, which make up the composite one.

On figure 3.3 you now know R and the angles θ and t (the apparent angular extent of the comet's tail). You want to know the length of the comet's tail in astronomical units. The answer may be found by construction. Draw on a sheet of graph paper a line to represent the line joining Sun to Earth on a suitable scale, for example, 100 mm to represent 1 AU. Mark the angles θ and t with a protractor. By trial, or by use of a compass, complete the large triangle so that the side SC, the line joining the Sun to the comet, is equal to R. Continue the line SC as in the diagram to where it intersects the second line from E. The intersection at T is the tip of the comet's tail. The linear length, s, of the tail is the distance CT in the diagram. Measure this distance with a ruler and convert to astronomical units according to the scale of the diagram.

Figure 3.5 The part of the sky which includes the Sun and the comet on a flat map. Right ascension (α) is measured eastwards (i.e. towards the left) along the celestial equator in hours (1 hour = 15°) and declination (δ) is measured vertically. In the absence of a globe this map may be used. Mark the positions of the Sun and the comet by their coordinates, and use the vertical scale to convert their separation (θ) to degrees.

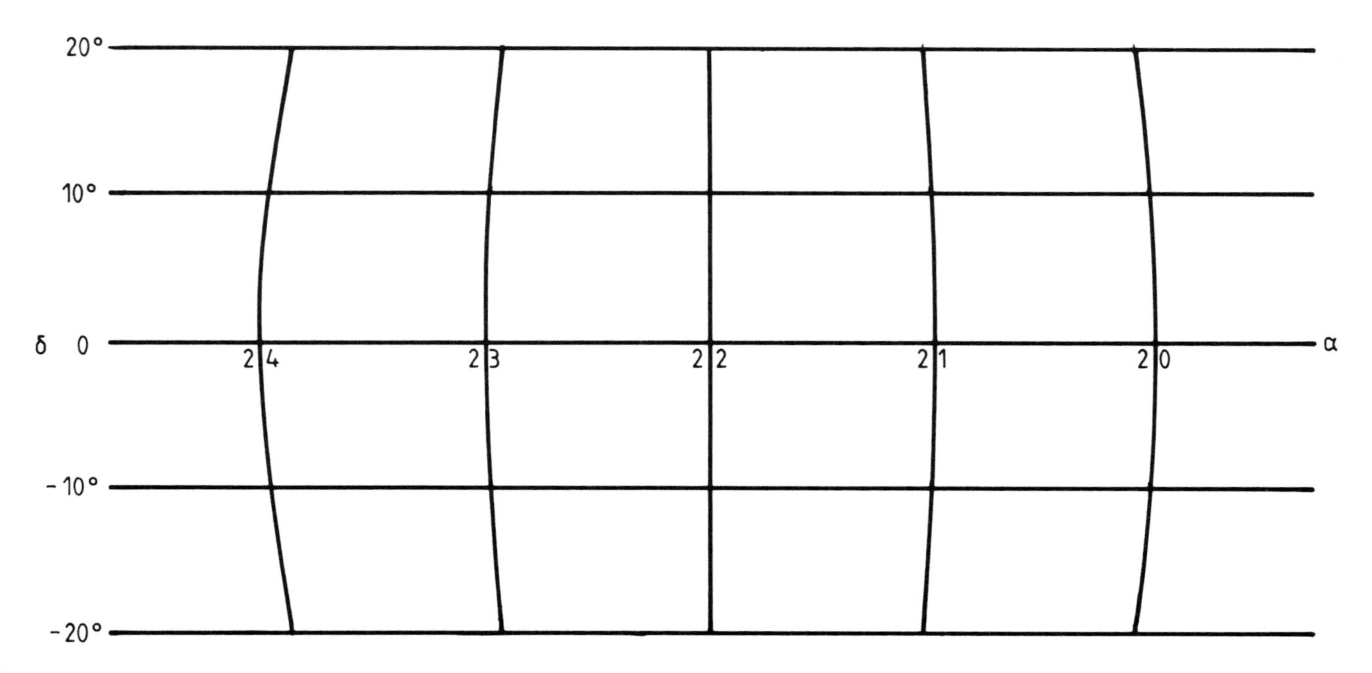

If you prefer, you may solve the triangle by trigonometry. By the sine rule in the smaller triangle in figure 3.3:

$$s = d \sin t / \sin (b - t)$$

where the angle b shown in the diagram must first be calculated by the sine rule in the large triangle. The distance d of the comet from the Earth is also found from the large triangle either by trigonometry or from the construction.

THE COMET'S MOTION AND CHANGES IN ITS STRUCTURE

Photographs of Halley's comet (plates 3.2 and 3.3) taken at different times show two features—the motion of the comet with respect to the stars, and physical changes in the tail. The photographs were taken at 18h 00m UT on March 9 and 10, i.e. exactly 24 h apart. The second photograph is part of the composite plate 3.1. The distance of the comet from the Sun on the two occasions was 0.84 AU on March 9 and 0.85 AU on March 10.

The straight ion tail, once formed, remains attached to the comet for some time but may abruptly break off and drift away with ever-increasing speed until it joins the main stream of the solar wind. The separation of a tail from the comet is called a *disconnection event*. The phenomenon did not escape the notice of observers in the early part of this century, but it was not understood and was largely forgotten. A spectacular disconnection event occurred in the tail of Halley's comet during its most recent appearance, followed by the outward movement of the tail which is seen on the pair of photographs taken on successive days.

Exercise 2. By comparing the photographs of the comet taken on March 9 and 10, calculate the rate of the comet's motion on the plane of the sky in metres per second.

Suggestions. Measure the motion by placing a transparent overlay on the first photograph and tracing on it the outline of the sharp front edges of the comet's head and the positions of a selection of stars around the head. Superimpose the overlay on the second photograph to make the star images around the head coincide (on account of the curvature of the sky, the images over the entire field will not coincide perfectly). Draw the comet's head in its new position. Measure the displacement in millimetres and convert it to angular measure and then into radians. You will be able to work out the scale of the photographs by comparing them with the composite photograph the scale of which has already been found in Exercise 1.

The linear distance travelled by the comet is the displacement in radians multiplied by the distance d to the comet. Strictly one should use the average d between March 9 and 10. However, d changes so little between the two dates that its value on March 10 (already calculated in Exercise 1) may be used in this instance as well. (You can satisfy yourself on this point by considering to what extent the large triangle in figure 3.3 changes between the two dates.) Convert the linear displacement to velocity in metres per second.

THE DISCONNECTION EVENT

The same photographs (plates 3.2 and 3.3) record the disconnection of an ion tail and its drift away from the comet. As the tail becomes disconnected another one forms. The motion of the disconnected tail may be followed by tracing the course of the distinctly recognisable knot seen at the breakaway point.

Exercise 3. Calculate the speed at which the disconnected tail separates from the main body of the comet.

Suggestions. The procedure is the same as in the previous exercise. Trace on a sheet of transparent overlay on the earlier photograph the sharp front of the comet's head and the position of the inner knot. Draw a line also showing the direction of the ion tail. Lay the overlay on the second photograph, lining up the front of the comet's head and the direction of the tail as carefully as possible, and mark the position of the knot. The direction of the tail shifts slightly from one photograph to the next because of the motion of the comet; there is also a small effect due to the curvature of the sky. You will therefore have to twist the overlay a little to get the tail lined up correctly the second time.

The overlay now shows the head and two successive positions of the knot. Measure in millimetres the distance of the knot from the head in the first photograph, and also the distance moved by the knot between the first and second photographs in millimetres. Convert these to degrees, from the scale of the photographs which you already know from the previous problem.

Convert the angular distance moved by the knot to true distance in space by making use of the small triangle in figure 3.3, as in Exercise 1, replacing the angle t by the angle travelled by the knot and increasing the angle θ by the angle from the head of the comet to the first position of the knot. The average speed of the knot is the distance moved divided by the time interval. The result should be given in metres per second in order to compare the speed with that of the solar wind, which is given in the Discussion.

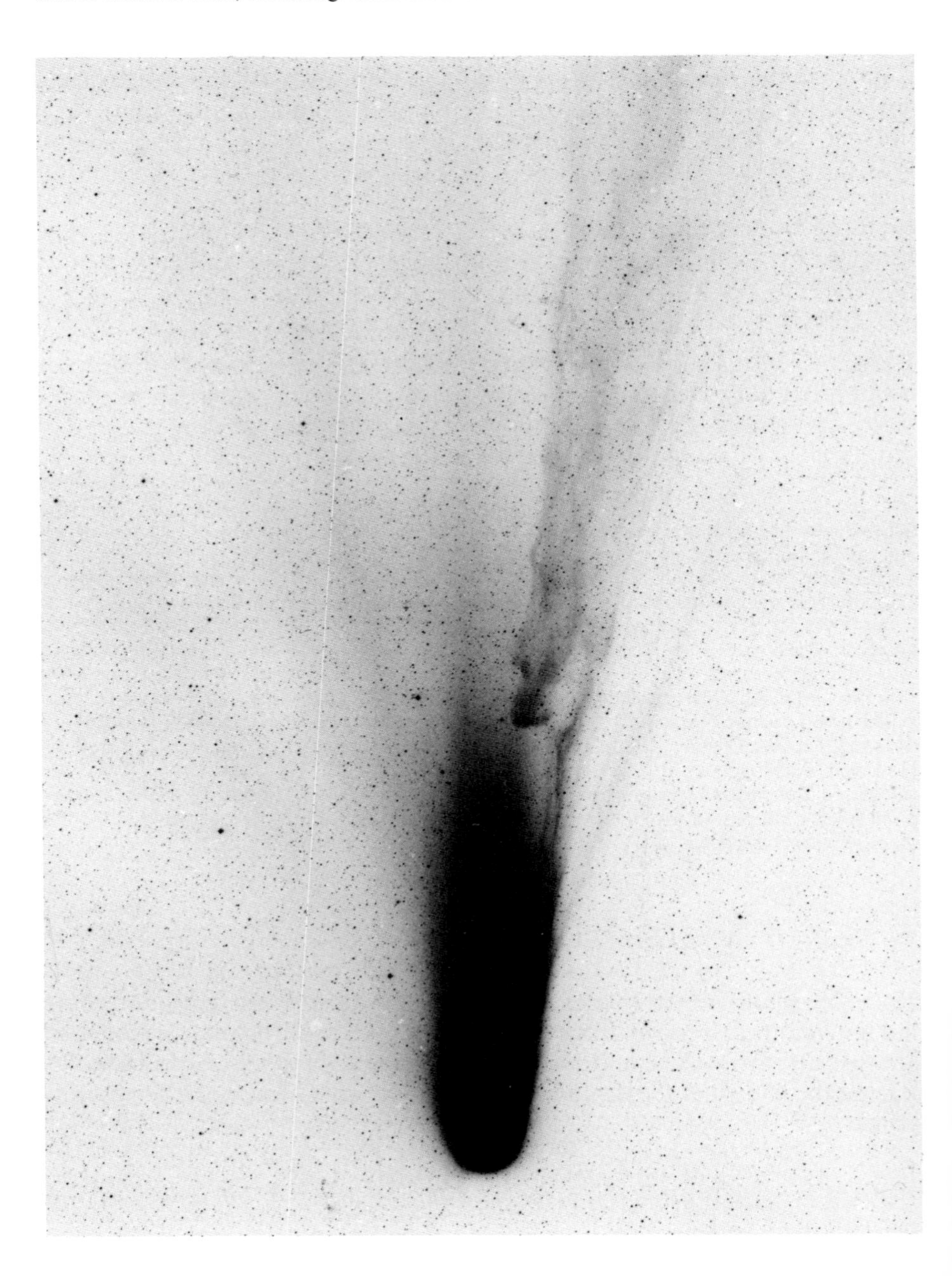

Plate 3.2 Halley's comet on March 9, 1986 (© Royal Observatory, Edinburgh).

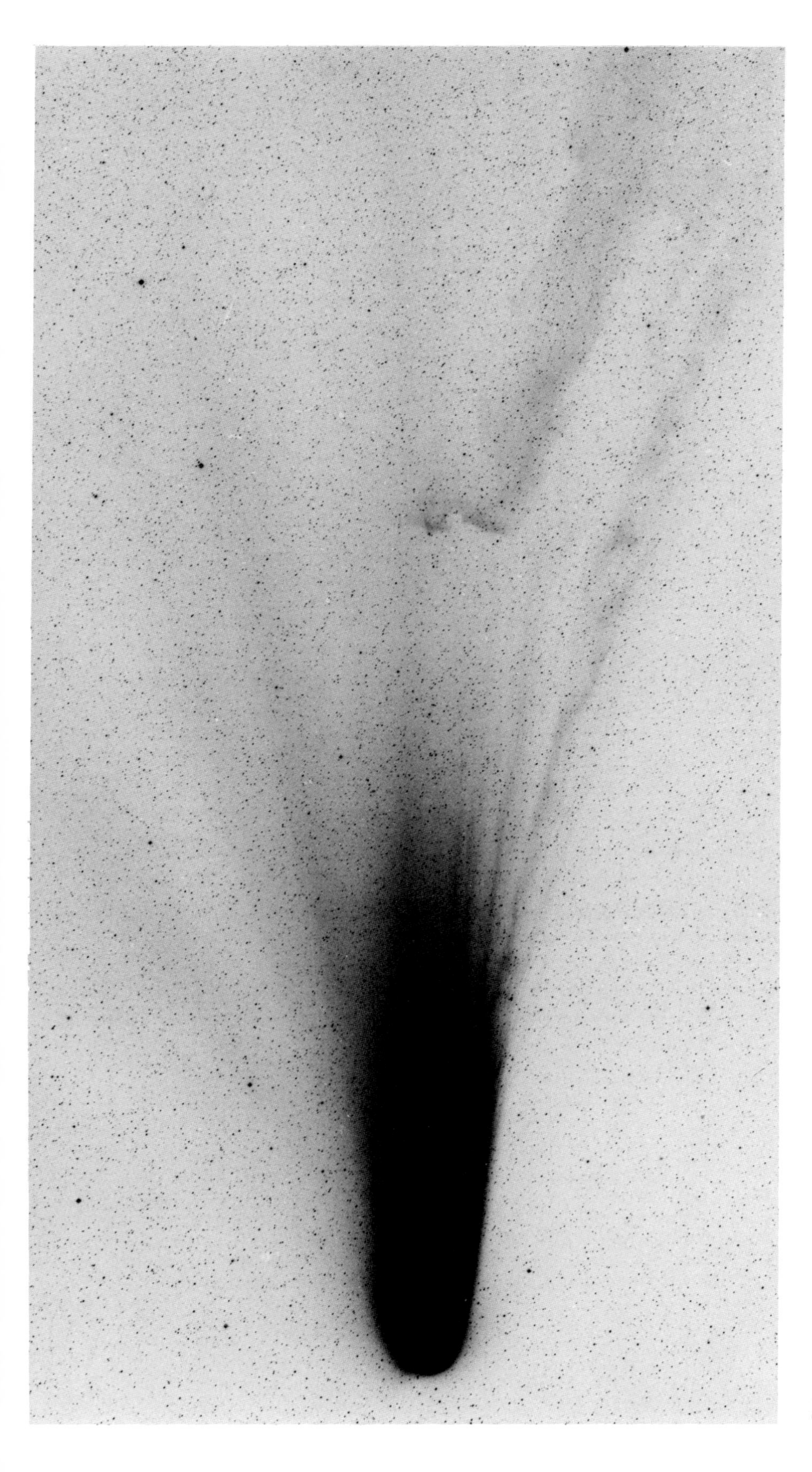

Plate 3.3 Halley's comet on March 10, 1986 (© Royal Observatory, Edinburgh).

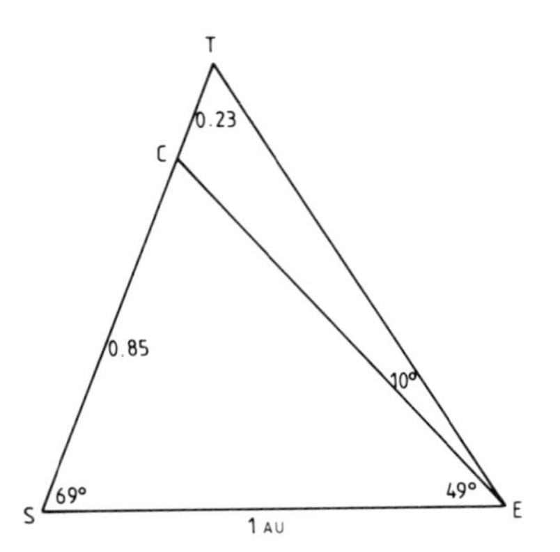

Figure 3.6 Actual construction of figure 3.3 drawn to scale.

SOLUTIONS AND DISCUSSION

1. The angle φ, measured from a globe, is 13°; it coincides with the direction of the ion tail nearest to the comet's head. This angle cannot be measured adequately from the map; however its exact value is not needed for the next calculation.

Figure 3.6 is the same as figure 3.3, drawn to scale. On this diagram
angle θ (Sun to comet) = 49°,
angle b (in triangle, by construction) = 62°,
d (measured in triangle) = 1.05 AU,
angle t (length of tail) = 10°,
linear length of tail by algebra = 0.23 AU.

2. Angular motion of comet in 24 h = 42 arcmin = 1.22×10^{-2} rad.
Distance travelled = angle × d = $1.22 \times 10^{-2} \times 1.05$ AU = 1.28×10^{-2} AU = 1.92×10^{9} m.

Divide this figure by 86 400 (number of seconds in 24 h) to get:

velocity = 22×10^{3} m s^{-1}

The following figures are given for comparison:

Earth's orbital velocity, 30×10^{3} m s^{-1}.
Comet's average orbital velocity on March 9–10, 47×10^{3} m s^{-1}.
Comet's velocity with respect to the Earth in the line of sight, 42×10^{3} m s^{-1} approaching.

3. Distance of knot from comet's head on first photograph = 1.60°, distance moved in 24 h = 1.94°.

The linear distance moved is calculated from the small triangle in figures 3.3 or 3.6, using slightly different angles. The knot is 1.6° farther out than the comet's head; the angle θ is larger and the angle b smaller by this amount than in Exercise 1, making $b = 60°$ (to the nearest degree) and slightly altering d. The angle which replaces t in the diagram becomes 1.94°.

By algebra, distance moved in 24 h = d sin 1.94°/sin 58° = 0.042 AU. Converting to metres per second as in Exercise 2, velocity = 73×10^{3} m s^{-1}. This is an average velocity; the tail is accelerating away from the Sun but has not yet merged with the local solar wind which, according to space observations made at the same time, was 35×10^{4} m s^{-1}.

Two comments are worth making with regard to calculations 2 and 3. First, it has been assumed that the distance d of the comet from the Sun did not change over the period covered by the two observations. In fact there *is* a change because the comet and the Earth are both in motion; however, it is a small change which has relatively little effect on the size and shape of the large triangle of figure 3.3. One reason for this is that the angle in the sky between the comet and the Sun (θ in the triangle) cannot change very much in 24 h because (i) the Sun moves only 1° per day on the ecliptic and (ii) the angle moved by the comet in the 24 h interval is less than a degree, as observed when the two successive photographs are compared. Another reason is that R, the distance of the comet from the Sun, also changes relatively little between the two dates. The exact values of the relevant quantities on both dates are shown in table 3.2.

The second comment refers to the solution of the small triangle in figure 3.3. where it should be noted that the small angle has to be measured with greater precision than the large angles.

Table 3.2

Date	R	d	θ	b
March 9, 18h	0.84	1.07	47.7°	61.3°
March 10, 18h	0.85	1.05	49.3°	62.2°

Observations of the disconnection event made at observatories around the globe, including the UK Schmidt Telescope photographs discussed here, provided an excellent record of its progress over more than 24 h. Analysis of the combined observations indicated that the tail broke away on March 8.39 (the time is given in decimals of a day) with a velocity of 24×10^{3} m s^{-1}, accelerating thereafter at 0.27 m s^{-2}. You may wish to try out how well your observations of the motion of the disconnected tail agrees with this analysis.

4 The Milky Way

Looking around the sky with the naked eye one finds that the stars are not scattered at random but are more numerous in some parts of the sky than in others. This is more obvious for the fainter stars which, seen through binoculars or a telescope, show a striking concentration towards the luminous belt in the heavens known as the Milky Way.

The Milky Way outlines a complete great circle around the sky. It has maximum brightness in the southern constellation of Sagittarius; in the northern hemisphere it includes the conspicuous constellations of Cygnus, Cassiopeia and Orion. The phenomenon of the Milky Way is produced by the projection on the sky of myriads of faint stars, members of a huge agglomeration of stars of which the Sun is one. In astronomical terminology the entire system is called the Galaxy from the Greek word for Milky Way, while the name 'Milky Way' is usually reserved for the visible starry belt. By analogy, external stellar systems are also called galaxies, our own home galaxy being distinguished by the title 'Galaxy' (with a capital G) or 'the Milky Way Galaxy'.

The present well attested model of the structure of the Galaxy was built up from observations of many kinds (figure 4.1). It is a system composed of stars, dust and gas, of mass 200 or more billion (thousand million) solar masses (2×10^{11}). It has a diameter of about 30 kiloparsecs (kpc), with the Sun at a distance of 8.5 kpc from the centre according to the best modern estimate, though 10 kpc had for a long time been the adopted distance. Three separate parts are recognised in the system's structure: the halo which is roughly spherical in shape, the disk which is flat, and the spiral arms which form an even flatter system within the disk. These subdivisions have different evolutionary histories. The halo contains the oldest stars in the Galaxy (10 to 20×10^9 years) coeval with the Universe. Though it occupies a very large volume the mass of the observed halo is only a small fraction of the total mass of the Galaxy. After the formation of the halo, the rotating Galaxy became flattened into a disk out of which later generations of stars, including the Sun, condensed. The disk contains most of the Galaxy's mass. The disk stars have ages ranging from 5 billion (5×10^9) years (which is the age of the Sun) downwards. The spiral arms are the regions where the very youngest stars, with ages less than 10^8 years, are found, and where star formation is still going on. The raw materials of stars are found abundantly in the spiral arms—dust, neutral hydrogen gas and large clouds of molecular gases. The various constituents of the Galaxy fall into one of two categories or *populations*. Population I includes objects associated with youth—young stars and also gas and dust out of which stars are formed. This population embraces both the spiral arms and the disk of the Galaxy (and of external galaxies). Population II includes older, evolved stars and star systems, specifically the globular star clusters. The difference in populations is explained theoretically in terms of age and evolution.

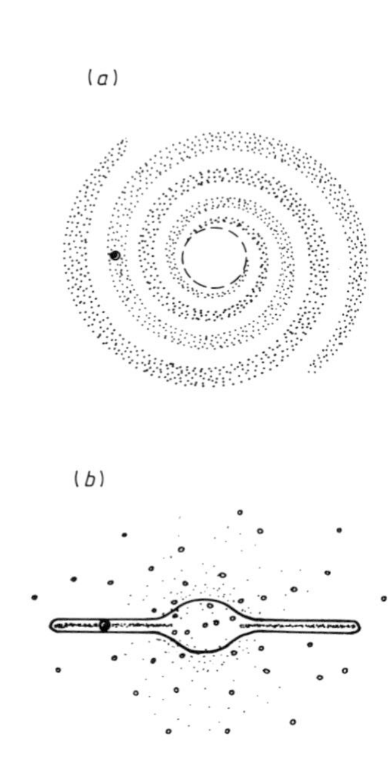

Figure 4.1 Model of the Galaxy, showing the position of the Sun at a distance from the centre (*a*) in the plane, (*b*) perpendicular to the plane showing globular clusters in the halo.

The great circle of the Milky Way in the sky defines the plane of the galactic disk, and its brightest regions point to the galactic centre. For studies of the structure of the Galaxy it is convenient to use the great circle of the Galaxy as an equator of the celestial sphere and the direction to the

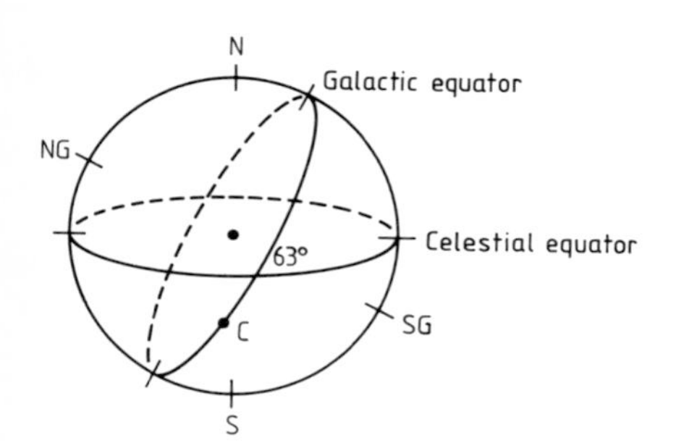

Figure 4.2 The plane of the Milky Way marked out by a great circle on the celestial sphere. C is the centre of the Galaxy, N and S are the north and south celestial poles and NG and SG are the north and south galactic poles.

centre of the Galaxy as a reference point on that circle. The galactic equator stands at an angle of 63° to the celestial equator and the galactic centre is at a point on the galactic equator 33° south of the intersection of the celestial and galactic equators (figure 4.2). The equatorial coordinates of the galactic centre are:

17h 45m (266°), −29° (in the constellation of Sagittarius).

Positions on the sky in the galactic system of coordinates are:

(i) longitude l, measured eastwards from 0° to 360° along the galactic equator, from the direction of the galactic centre, and
(ii) latitude b, measured from 0° to +90° north and −90° south of the galactic equator.

The photographs (plates 4.1, 4.2 and 4.3) show three areas of the sky at galactic latitudes −90° (the south galactic pole), −53° and −20°. The photographs, all taken with the same instrument under similar conditions, obviously confirm the increase in star densities as one goes from the galactic pole toward the galactic equator. Another photograph later in this book (plate 8.1) shows a field at galactic latitude 0° (the galactic plane) where the stars are even more numerous. The galactic pole field happens to include a conspicuous external galaxy, which should be ignored in this exercise.

The first systematic study of the distribution of stars in the stellar system was made by William Herschel at the end of the 18th century. Herschel attempted, without success, to find the distances to the stars. This meant that he could not tell the scale of the stellar system. However, he was able to construct a model of the distribution of stars in space, based on star counts which he made in 5000 different directions in the sky. Herschel assumed that all stars were more or less uniformly distributed in space and also that his telescope reached the limit of the system. On these assumptions it followed that the directions where stars were most numerous indicated the farthest extensions of the system. The result was the famous 'millstone' model of a flattened system with the Sun at or near its centre.

In Herschel's model the number of stars per unit volume of space is the same throughout the system. The observer's field of view is a cone with the observer at the apex (figure 4.3(a)) or a pyramid in the case of a rectangular field on a photograph (figure 4.3(b)). The volume of a cone or pyramid is proportional to the cube of its radius, the radius in this case being the distance to the edge of the system. The number of stars is proportional to the volume of space observed, which is proportional to the eube of the distance to the edge of the system. Conversely, the distance to the edge of the system in a particular direction is proportional to the cube root of the number of stars observed in that direction.

Later, when the distances of some stars had been measured and when it was known that there exists a wide range of luminosities among stars, the method of star counting was resumed as a means of discovering the shape and size of the stellar system. Instead of counting all stars, as Herschel had done, astronomers now sorted the stars by magnitude and counted the numbers in various magnitude intervals. (The magnitude system is described in Appendix 2.) Star counting remains an important tool of the astronomer to the present time.

Early results of counting by magnitude contradicted Herschel's model of a system of stars uniformly spread in space throughout the Galaxy. Instead, the density of stars in space appeared to thin out with distance from the Sun. This conclusion was reached as follows. If all stars are equally luminous, then by the inverse square law their apparent brightness decreases inversely as the square of the distance. At the same time, if the density (the

Facing page:
Plate 4.1 Photograph of a field near the south galactic pole of the sky, showing standard stars. The large galaxy is NGC 253 which happens to fall in the field (© Royal Observatory, Edinburgh).

Page 26:
Plate 4.2 Photograph of a field at galactic latitude −53° showing standard stars (© Royal Observatory, Edinburgh).

Page 27:
Plate 4.3 Photograph of a field at galactic latitude −20° showing standard stars (© Royal Observatory, Edinburgh).

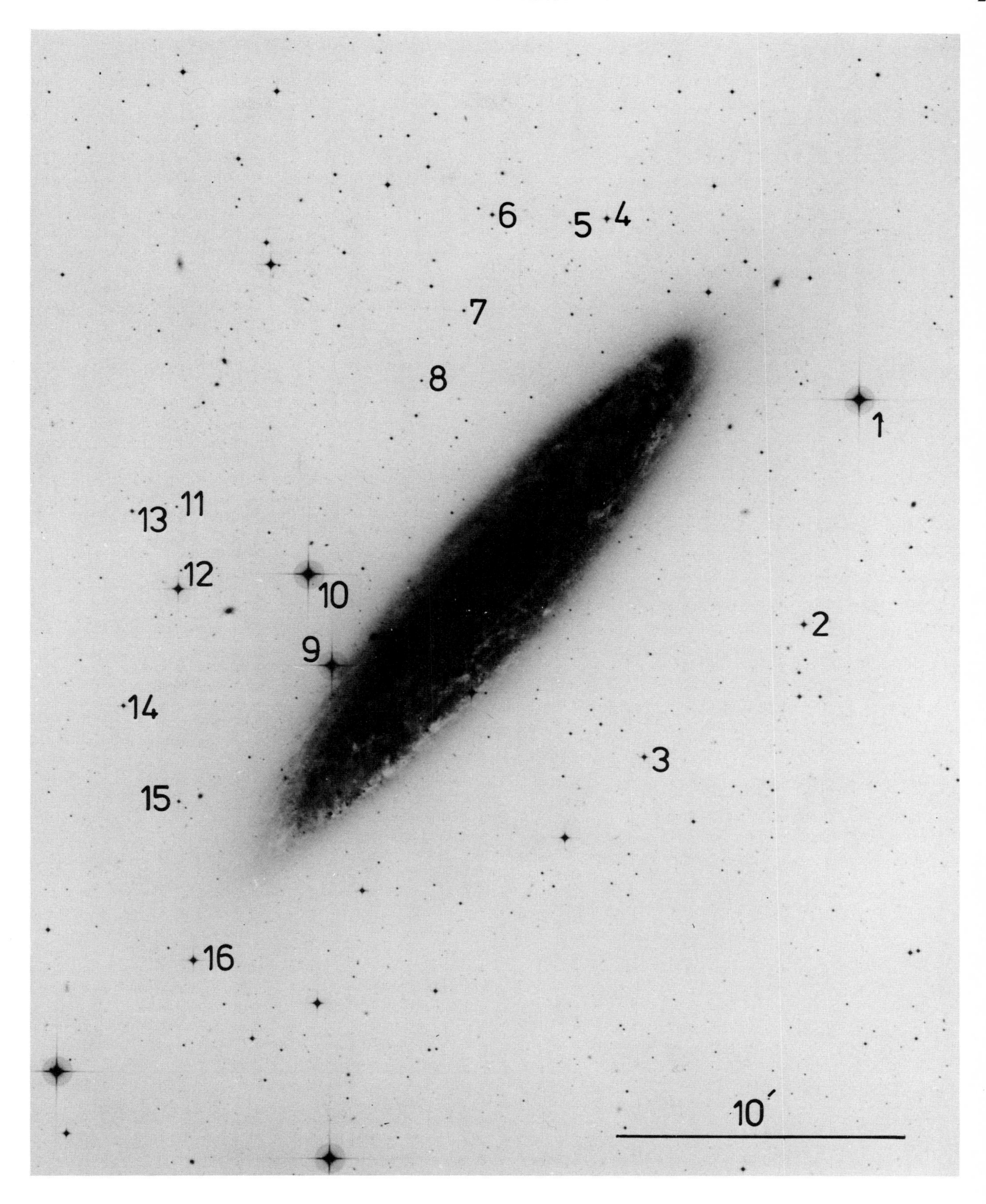

Plate 4.1

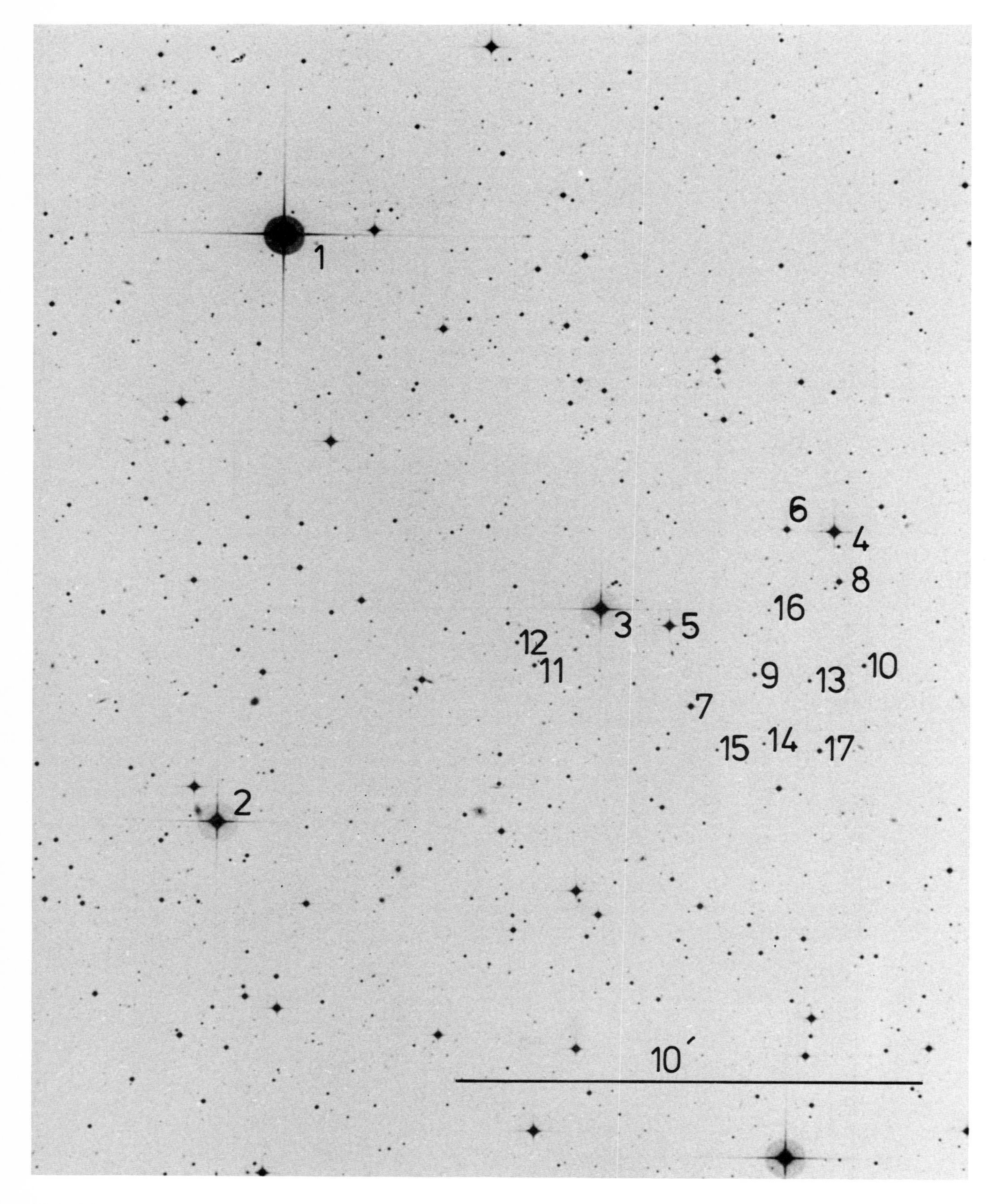

Plate 4.2

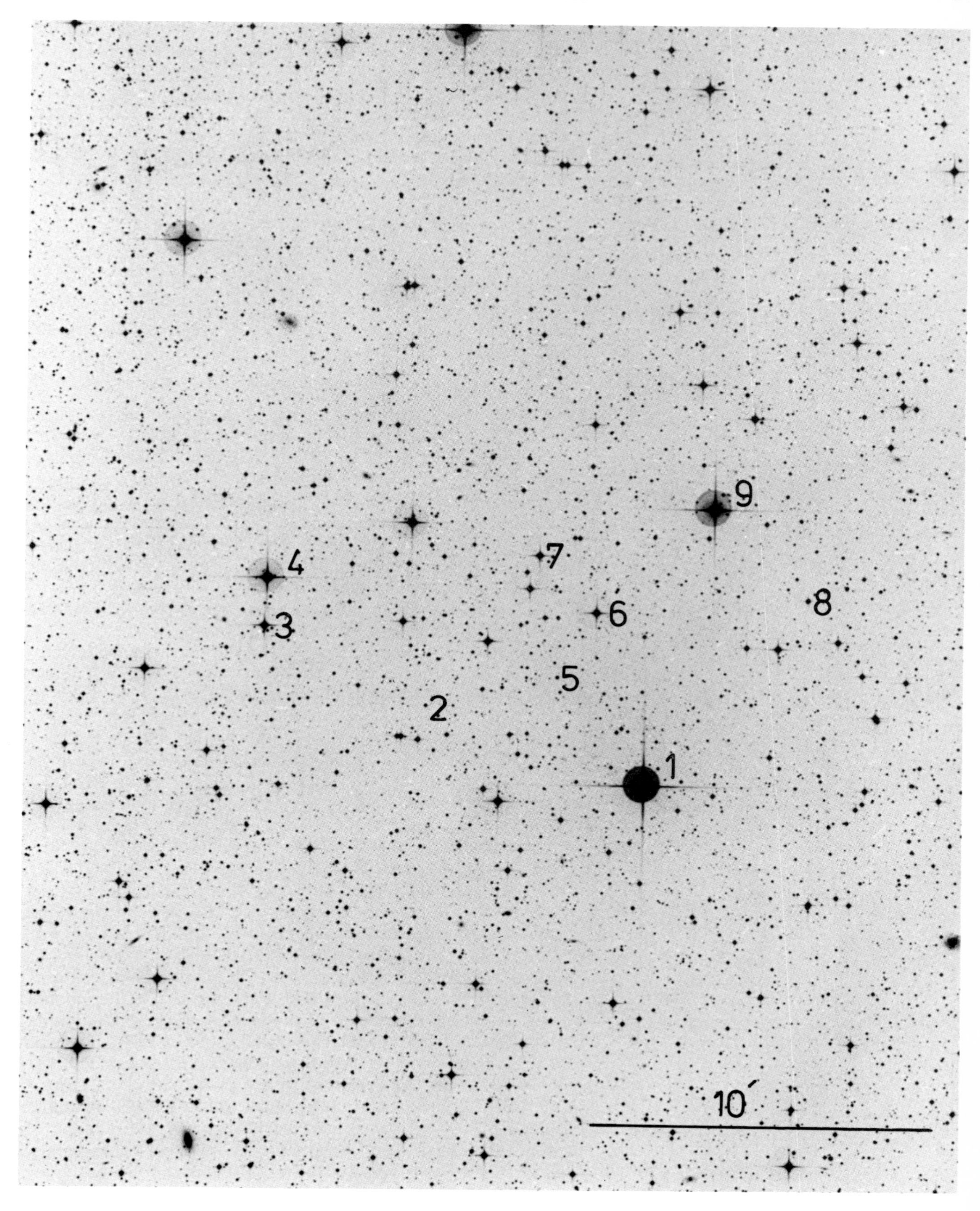

Plate 4.3

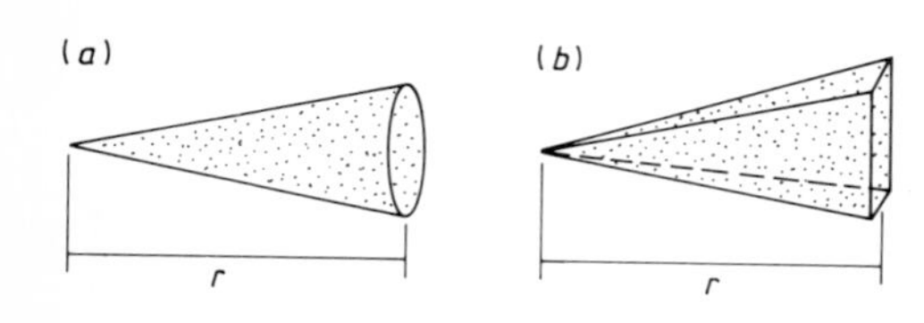

Figure 4.3 (*a*) Cone or (*b*) pyramid of stars in uniformly filled space, seen by an observer at the apex. In each case the number of stars is proportional to the cube of the depth r.

number per unit volume of space) of stars is the same to the limit of the system, then, as Herschel deduced, their total number increases with the volume of space observed.

If m is the limiting magnitude, B the apparent brightness of stars of magnitude m, r the distance to which the stars are counted and N the total number of those stars; and if B_1, r_1, and N_1 are the same quantities for limiting magnitude $(m+1)$, then by the inverse square law

$$B/B_1=(r_1/r)^2$$

and the ratio of the volumes occupied is

$$(r_1/r)^3=N_1/N.$$

Eliminating r,

$$B/B_1=(N_1/N)^{3/2}$$

or in logarithms

$$\log(B/B_1)=\tfrac{3}{2}\log(N_1/N).$$

By definition of magnitude, $2.5 \log(B/B_1)$ is the magnitude difference between two stars of apparent brightnesses B and B_1, in this case 1 magnitude. Therefore $\log(N_1/N)=0.6$ for a difference of 1 magnitude.

This result is true whether or not all stars are equally luminous, provided the proportion of stars of different luminosity is the same throughout the space observed. An increase of 0.6 in $\log N$ is equivalent to an increase in N by a factor of four per magnitude. This means that for each step of 1 magnitude included in the counts, the number of stars should increase by a factor of four if the density of stars were the same everywhere.

In fact, observations, in whatever direction of the sky, always showed a variation of less than 0.6 in $\log N$ per magnitude.

The exercise, in which fields in three different galactic latitudes are examined, demonstrates how this result was derived and illustrates the general principle and method of star counting.

PHOTOGRAPHIC STELLAR PHOTOMETRY

Before beginning the counts it is essential to know the magnitudes of the stars in question. The images of bright stars on the photographs are larger and blacker than those of faint ones, but the relation between the size of an image and the brightness of the star is not a linear one in all cases. Ideally, star images are all extremely small; however, the images actually formed depend on various instrumental factors and on the quality of the atmosphere through which the light has passed. It also depends on the particular photographic emulsion used. On striking the sensitive emulsion, photons of light are scattered causing the grains in the emulsion to blacken after development; the more light arrives in a star image, the more grains are blackened. In the case of very faint images, only a few grains become blackened; in the case of bright stars, the image on the photograph extends over a large area and its central parts are completely black. Very bright images on our photographs have spikes in the form of a cross; these are caused in reflecting instruments by diffraction of light by obstruction in the telescope tube (the plateholder in the case of the Schmidt Telescope photographs shown here). Bright images on these photographs are also surrounded by a halo, which is another instrumental effect.

To relate the dimensions and blackness of a star image to the light flux from the star one must calibrate the photograph by observing a representative set of stars—the standard stars—which have previously had their magnitudes measured independently. The usual method is by the use of a photoelectric photometer which has a linear response to light.

A set of standard stars is marked on each photograph. The photoelectric magnitudes of these stars are given in tables 4.1, 4.2 and 4.3. The photographs (on J emulsion with a suitable blue filter) correspond closely to the standard B band (Appendix 2); the appropriate magnitudes given in the tables are therefore B magnitudes. A calibration curve is set up by plotting the diameters of images of the standard stars against their magnitudes, and drawing the best smooth curve through the points. The magnitude of any other star in the field may then be found by reading from the graph the magnitude corresponding to its image diameter. Photoelectric magnitudes are correct to 0.01 or even 0.005 magnitude (an accuracy of 0.01 magnitude corresponds to an accuracy of 1% in the measured light flux). Photographic magnitudes measured on original glass negatives are somewhat less accurate (0.03 mag). The amount of information is much reduced on printed copies of star images such as are reproduced here.

Exercise 1. Draw calibration curves for Fields 1, 2 and 3 (plates 4.1, 4.2 and 4.3) by plotting the diameters of the images of the standard stars against their magnitudes. The exercise is intended to show that the curves are not linear and that they may vary from one photograph to another.

Measure image diameters to 0.1 mm. Begin with plate 4.2, which is the one discussed in the Solutions. Compare your calibration curve with that shown in the Solutions before proceeding to Exercise 2.

Exercise 2. Investigate the rate at which numbers of stars increase with magnitude in various galactic latitudes. Give the result in the form of an increase in the logarithm of the number per magnitude. Note that it is necessary to count only to two suitable magnitude limits in order to get the rate of increase.

Table 4.1 Standards in Field 1 (plate 4.1). ($b = -90°$.)

Star	B
1	9.66
2	13.22
3	12.42
4	13.48
5	17.29
6	14.64
7	16.06
8	16.65
9	9.90
10	9.45
11	17.18
12	11.31
13	14.90
14	14.40
15	15.64
16	12.29

Table 4.2 Standards in Field 2 (plate 4.2). ($b = -53°$.)

Star	B
1	7.85
2	10.35
3	10.35
4	11.82
5	12.72
6	14.50
7	14.63
8	14.80
9	16.55
10	16.55
11	16.52
12	17.35
13	17.48
14	18.79
15	18.45
16	19.26
17	20.62

Table 4.3 Standards in Field 3 (plate 4.3). ($b = -20°$.)

Star	B
1	8.29
2	16.10
3	11.60
4	10.08
5	16.05
6	12.59
7	12.32
8	14.07
9	9.41

It does not matter which field you begin with, though the galactic pole region is perhaps the easiest because it has fewest stars. (The area occupied by the galaxy image on this photograph should, however, be excluded from the counts.) Study the calibration curve and choose two magnitude limits for star counting. The brighter limit should be such as to provide a reasonable number of images—say 30 or 40—in the field; the fainter limit should provide considerably more counts but should not include the very small images which give inaccurate magnitudes. It is easier to fix limits of image diameters which are whole numbers of the units of measurement than limits which correspond to exact whole numbers in magnitudes. Bright images have diffraction spikes whose lengths may be found useful as calibrators. The presence or absence of spikes is one way of fixing a magnitude threshold.

Suggestions. Count the bright images first. Scan the photograph in strips and mark (either in pencil lightly on the photograph or on a transparent overlay) all images which are as large as, or larger than, the chosen upper limit; this ensures that each image is counted once and once only. The numerous counts to the fainter limit could be done in small squares. Cut out a square window (1 cm × 1 cm, for example) from a piece of paper and move it systematically through the field, writing down the number you count at each position. If the stars are very numerous it is enough to count some sample squares and scale up the counts to give the total number for the whole field. This number will, of course, include the large images already counted. Take care to avoid images of galaxies; the faint ones are distinguished from stars by their non-circular shape.

Calculate the numbers per unit area to each magnitude limit and take their logarithms. Finally, subtract the logarithms and divide by the magnitude difference between the limits to give the logarithmic rate of increase of star numbers with magnitude.

Repeat the procedure for the other fields. It does not matter if the chosen magnitude limits differ from one field to another, or if areas of different sizes are counted on the different fields, because, for the moment, we are interested only in the rate at which the numbers increase, not in the numbers themselves.

Exercise 3. Using the result of the previous exercise and taking into account the scale marked on each photograph, calculate the logarithm of the number of stars per square degree at the various galactic latitudes to limiting magnitude 15.

Suggestions. The logarithms of the numbers to two different magnitude limits are known from the previous exercise. The logarithm of the number to a specific magnitude between the two limits is deduced by interpolation (that is, by calculation)—it is not necessary to count anew in order to get the answer. The area of each field in square arcminutes (converted to square degrees by dividing by 3600) is easily found by using the scale marked on each photograph. In the case of Field 1 (plate 4.1), the area covered by the large galaxy image should be excluded.

SOLUTIONS AND DISCUSSION

1. Figure 4.4 is the calibration curve of Field 2 (plate 4.2). The diameters of the images were measured with a graticule with 0.1 mm divisions, viewed through a 10 × magnifier, and recorded to the nearest division. It is seen that the curve is not a straight line; it is steep at the top end and tends to become flat at the bottom end. This shape is characteristic of photographic calibration curves. The most reliable part of the curve is the almost straight middle part (in this case in the magnitude range between about 10 and 16 or 17). At the bottom end of the curve (shown dotted) which involves the faintest stars a small difference in image diameter produces a large change in magnitude, and the faintest images (in this case from magnitude 18 onwards) are effectively all of the same size. (On the original photographs this levelling off happens much later; accurate magnitudes are actually achieved for stars as faint as 21 mag.)

When comparing your calibration curve with figure 4.4, remember that the measurements plotted here were made on a photographic print which was subsequently printed on a different scale, and also that there are personal random errors of measurement. Do not expect, therefore, to reproduce exactly the same curve. The shape, however, ought to be similar.

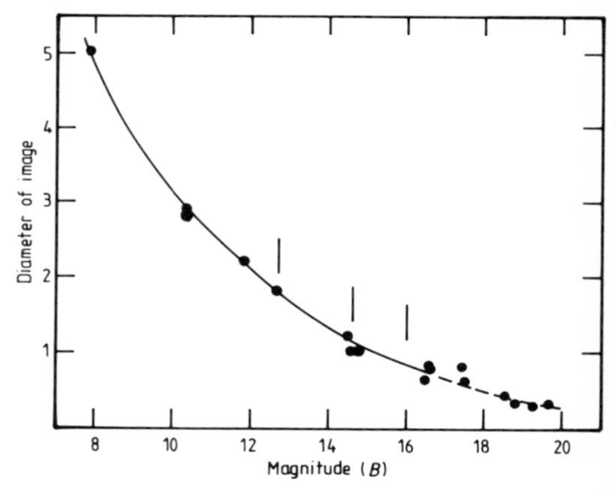

Figure 4.4 Calibration curve for Field 2. Diameters of star images in arbitrary units are plotted against magnitudes.

2. Three magnitude limits are marked on figure 4.4. The brightest one, which was first tried, produced only 13 images which was considered insufficient for the purpose. The second limit is that at which spikes are just visible on the images. By measuring the diameter of a number of those images the corresponding magnitude, read off the graph, turned out to be 14.6. It is easy to recognise images which have spikes, and so it was easy to count the total number of stars down to this limit, which was 46. The third limit had to be well above the flat part of the calibration curve. The limit of 16.0 mag was chosen. The counting to this limit required the careful measurement of the dimensions of borderline images. The total number in this category was 153.

The rate of increase in star numbers with magnitude is therefore

$$\log (153/46)/(16.0-14.6)=0.52/1.4=0.37 \text{ per magnitude}$$

The probable error in a number counted in a random sample (such as this, because we would expect any neighbouring nearby field to give the same result, with only statistical variations) is the square root of the number. If the number is N, it is written with its probable error as $N \pm \sqrt{N}$. The two counts with their probable errors are therefore 153 ± 12 and 46 ± 7.

When the statistical error, and a possible error of about ± 0.3 mag in the magnitude range, are taken into account, the upper value of the result is still well below 0.6 per magnitude.

The other fields, for which it is not necessary to repeat details, yield similar results. Even in the low galactic latitude field, the rate also comes out at less than 0.6.

At one time this result was mistakenly interpreted as indicating a falling-off of star density with distance from the Sun in all directions and, indirectly, as indicating that the Sun is the centre of a system. The flaw in the model was the tacit assumption that space between the stars is completely empty and that light from all stars reaches us unimpeded. In fact, interstellar space in the galactic plane contains dust which dims and may even totally obscure the stars which lie beyond it. The discovery of interstellar dust was one of the factors which led to the revised model of the Galaxy, in which the Sun is no longer at the centre and in which the stars in the direction of the galactic centre increase with distance. A demonstration of the effects of interstellar obscuration is given in §8.

3. The area of Field 2 (plate 4.2) is 0.152 square degrees.

The numbers of stars per square degree to the magnitude limits 14.6 and 16.0 (from the numbers in the previous solution) are 303 and 1007 respectively; their logarithms are 2.48 and 3.00. The variation in the logarithm (0.37 per magnitude) is 0.15 for 0.4 magnitude; this is added to 2.48 (value for 14.6 mag) to give the value, 2.63, for 15.0 mag.

Table 4.4 is a list of adopted data on star numbers for comparison with your results. When comparing your results with these data remember that these are mean values over all galactic longitudes.

Table 4.4 Logarithm of the number of stars brighter than magnitude $m(B)$ per square degree for different galactic latitudes.

Magnitude m	± latitude			
	0°	20°	50°	90°
10	0.97	0.66	0.40	0.27
11	1.43	1.08	0.80	0.66
12	1.88	1.50	1.19	1.03
13	2.30	1.90	1.54	1.39
14	2.72	2.28	1.88	1.71
15	3.12	2.65	2.20	1.97
16	3.48	3.00	2.48	2.24

5 Stars in Motion

Stars move in space. Their motions in the plane of the sky are recognised as actual displacements, while motions in the line of sight show up in the form of Doppler shift in their spectra.

PROPER MOTION

It has been known since the 18th century that the stars are not truly fixed in space. The positions of certain stars, when re-observed after an interval of time, are found to change with respect to their neighbours. In fact all stars move in space, but their great distances make their apparent motion in the plane of the sky difficult to detect in the majority of cases.

Such motion is known as *proper motion*. It is usually given the symbol μ and is measured in arcseconds per year. The unit in which stellar distances are measured is the parsec (pc), defined as the distance at which 1 AU subtends an angle of 1 arcsec. An angular separation of 1 arcsec at a distance of 1 pc is thus equivalent to a linear separation in the plane of the sky of 1 AU. An angle of μ arcsec at a distance of r pc (figure 5.1) corresponds to a separation of $r\mu$ AU; proper motion of μ arcsec per year of a star at a distance of r pc corresponds therefore to a velocity of $r\mu$ AU per year.

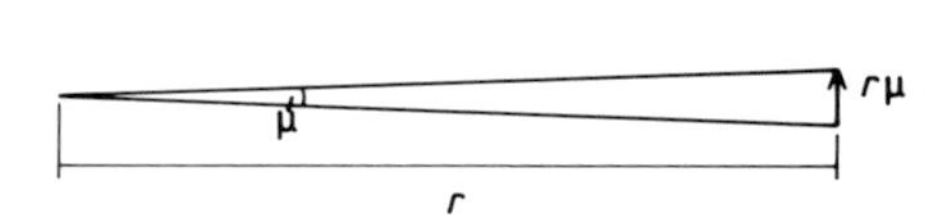

Figure 5.1 The relation between proper motion μ and transverse velocity for a star at a distance r.

$$1\ \text{AU} = 1.5 \times 10^{11}\ \text{m approximately}$$

$$1\ \text{year} = 3 \times 10^{7}\ \text{s approximately.}$$

Therefore 1 AU per year $= 5 \times 10^3\ \text{m s}^{-1}$ approximately. When exact values are substituted (Appendix 1) the result is $4.74 \times 10^3\ \text{m s}^{-1}$. The formula for the linear velocity at right angles to the line of sight of an object at distance r pc which has proper motion μ is then:

$$U = 4.74\ \mu r \times 10^3\ \text{m s}^{-1}.$$

The velocity U—called *transverse velocity* because it is across the sky—cannot be calculated unless the distance r of the star is known. Proper motion is more easily observed than parallax, and the number of stars for which proper motion is known is greater than the number for which distance is known. (Annual parallax is the change in a star's position relative to the background of distant stars due to the Earth's annual motion round the Sun, from which the star's distance is measured in terms of the dimensions of the Earth's orbit.) Improved methods of making precise observations of star positions (which now include observations from outer space) continue to add to the numbers of stars for which distances and motions are available. Motions of stars in the Galaxy are of the order of $3 \times 10^4\ \text{m s}^{-1}$, though some are as high as $10^5\ \text{m s}^{-1}$. The Sun's motion through space with respect to the neighbouring stars is $2 \times 10^4\ \text{m s}^{-1}$.

Exercise 1. The two photographs on plate 5.1 show the same area of sky, the second taken 10 years after the first. The field contains a star of observable

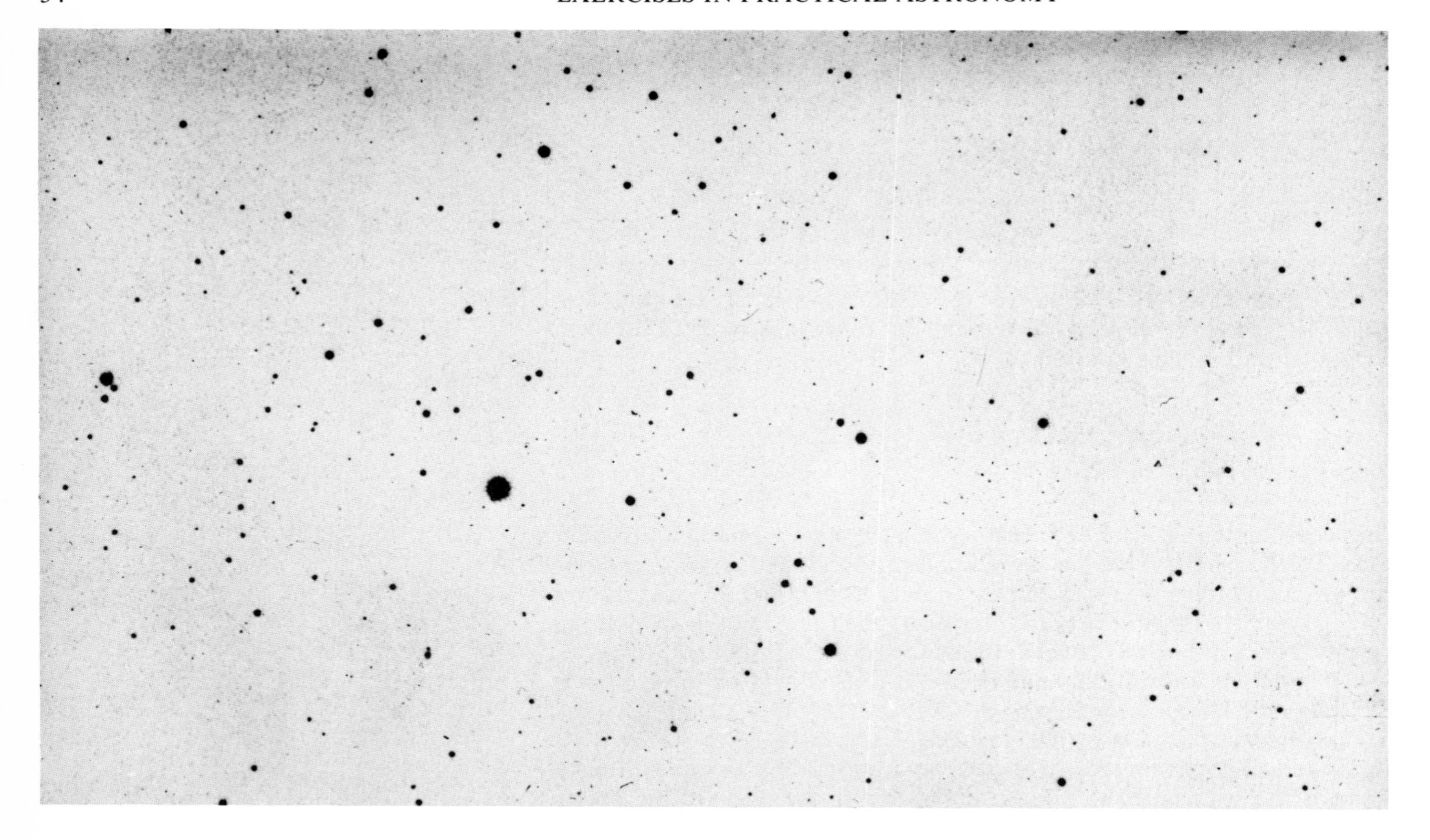

Plate 5.1

proper motion. Search for this star and find its proper motion in arcseconds per year (correct to 0.1 arcsec). Given that the distance to this star is 1.8 pc calculate the transverse velocity.

The classical method of discovering proper motion is to compare the two photographs by viewing them alternately in an instrument called a 'blink comparator'. The discordant jumping of the image of a moving star shows up against the unchanging positions of the other stars in the field. In the present instance the discordant star will have to be found by visual scanning of the photographs.

Suggestions. Having found the star, place a transparent overlay on one photograph and mark with a fine marker the positions of a number of stars in the field including the moving one. Transfer the overlay to the second photograph to match again with the positions of the stars. The shift in the position of the moving star should be measured with a millimetre ruler or graticule to the nearest quarter of a millimetre. Convert to arcseconds using the scale given in the caption of the photograph, divide by 10 (years) and apply the formula for transverse velocity in metres per second.

RADIAL VELOCITY

Proper motion denotes motion at right angles to the line of sight. Motion in the line of sight is observable from a displacement caused by the Doppler effect in the spectrum of a star. According to the Doppler principle a change in wavelength occurs when a source of light moves in the line of sight relative to the observer. The change in wavelength is proportional to the relative velocity v in the line of sight according to the formula:

$$(\lambda' - \lambda)/\lambda = v/c$$

where λ is the normal wavelength observed when there is no relative motion, λ' is the changed wavelength and c is the speed of light. It is usual to use the notation $\Delta\lambda$ for the change in wavelength, the symbol Δ meaning a small difference. The formula then is

$$\Delta\lambda/\lambda = v/c.$$

When the source is receding from the observer, v is positive and the change in wavelength is positive (i.e. wavelength increases); when the source is approaching, v is negative and the change in wavelength is negative (i.e. wavelength decreases). The relative velocity in the line of sight is known as the *radial velocity*. Wavelength of light increases along the spectrum from blue to red, and Doppler displacements are often described as being towards the red or towards the blue.

The Doppler formula does not involve the distance of the star; the radial velocity is obtained from the spectrum, irrespective of the distance of the source. For this reason the velocities of distant galaxies are observable while proper motion is detectable only for relatively nearby stars in the Galaxy.

Plate 5.2 shows the spectrum of a star (of spectral type A) together with the spectrum of a laboratory source of light (in this case iron) photographed simultaneously, from which any shift in the spectrum of the star may be looked for. The spectra are taken in the conventional way, that is, the spectrum of the star is in the middle horizontal strip of the photograph and the spectrum of the comparison laboratory source is divided vertically in two parts, so that by joining up the lines of the two parts a wavelength scale for the spectrum of the star in the central strip is provided. The lines in

Facing page:
Plate 5.1 Two photographs of the same region of the sky taken 10 years apart. The horizontal dimension of the photograph is 40.5 arcminutes (© Lick Observatory).

the stellar spectrum are absorption lines, whereas those in the comparison spectrum are emission lines. The spectrum of iron is particularly useful as a comparison because it contains hundreds of lines with very accurately known wavelengths. Many stars show iron absorption lines in their spectra; in stars of this type the Doppler shift is the displacement of the star's own iron lines with respect to the laboratory ones. However, as in the present example, the lines in the stellar and the laboratory spectra may be different. The lines in the stellar spectra are recognised from their relative positions; for example, the spectrum of an A type star is dominated by the series of lines of hydrogen (the Balmer lines) which have an unmistakable pattern. The photograph, of the violet end of the spectrum, includes four Balmer lines and also the easily identified pair of lines (called H and K) due to ionised calcium (Ca II). Though the measurement of radial velocity is very simple in principle, it is less so in practice because the shift in the lines is usually very small. In the example shown here, a star with a very large radial velocity has been chosen which shows a measurable Doppler shift on the enlargement.

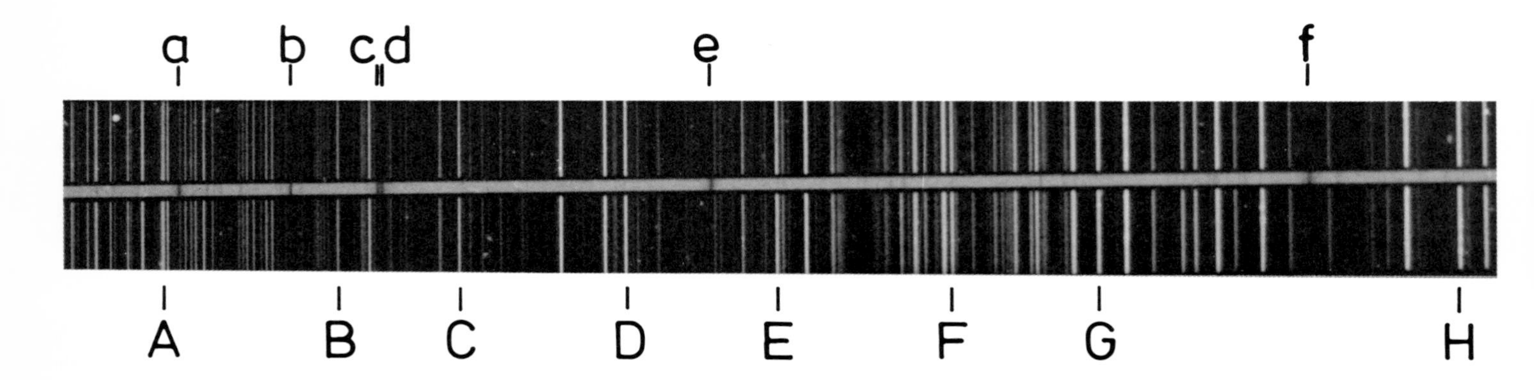

Plate 5.2 Spectrum of an A type star and comparison spectra of iron. A selection of stellar and iron lines are identified (© Observatoire de Marseille).

Exercise 2. Plate 5.2 is the spectrum of an A type star together with the spectrum of a laboratory source of light (iron). The 'rest' wavelengths of the lines in the spectrum of the star (that is, the wavelengths for a source which is not in motion) are listed in table 5.1. A selection of lines in the iron spectrum are marked on the photograph and their wavelengths listed in table 5.2. Calculate the radial velocity of the star, and decide whether the star is approaching or receding.

Table 5.1 Wavelengths of lines in the spectrum of the star. (Lines c and d are very close together.)

Line	λ (nm)	
a	388.90	Hydrogen
b	393.38	Calcium II
c	396.86	Calcium II
d	397.01	Hydrogen
e	410.17	Hydrogen
f	434.05	Hydrogen

Table 5.2 Wavelengths of lines in the spectrum of iron.

Line	λ (nm)
A	388.71
B	395.67
C	400.52
D	407.17
E	413.21
F	420.20
G	426.05
H	440.48

Suggestions. The obvious first step which you will consider is to measure the positions of the iron lines from some fixed point (e.g. the edge of the spectrum or the first iron line at the left end of the spectrum) and to plot a graph of wavelength against position. The result is a straight line. The positions of the stellar lines are measured from the same point, and the wavelengths corresponding to those positions are read off the graph. The differences between those wavelengths and the wavelengths in table 5.1 are the Doppler shifts. Though this method is the right one in principle, you will find that you may not be able to read wavelengths as accurately as you would like from the graph. A better way is to calculate the wavelengths of the stellar lines by interpolation between the iron lines. If λ_1 and λ_2 are the wavelengths of two iron lines which are x millimetres apart, and λ_3 is the wavelength of a stellar line which falls between them at a distance y millimetres from the first iron line, then the stellar line is a fraction y/x of the distance between the two iron lines, and its wavelength is therefore

$$\lambda_3 = \lambda_1 + (y/x)(\lambda_2 - \lambda_1).$$

The distances x and y should be measured as accurately as possible because the precision of the calculated wavelength depends on the smallest distance which you can measure. Try using a millimetre ruler to estimate positions to the nearest quarter of a millimetre. Repeat this procedure for each of the stellar lines. Draw up a list of $\Delta\lambda$ and λ; divide each $\Delta\lambda$ by its λ and multiply the result by the velocity of light (3.0×10^8 m s^{-1}) to get the radial velocity in metres per second. Take the mean. Remember to take into account the sign of $\Delta\lambda$, in order to decide if the star is approaching or receding.

SOLUTIONS AND DISCUSSION

1. The displacement (measured by one observer) is equivalent to 97 arcsec and was measured to 1 part in 10 on the photograph. The proper motion is therefore 9.7 ± 1 arcsec. The transverse motion is $4.74 \times 1.8 \times 9.7 \times 10^3$ m s^{-1} $= 83 \times 10^3$ m s^{-1}; this also has an uncertainty of 10%, so it is given as 83 ± 9 m s^{-1}. Your results should not necessarily be exactly the same as these figures, but should fall within the error limits.

The star is Barnard's star which has the highest known proper motion. The exact value of its proper motion is 10.31 arcsec.

By definition, the distance of a star in parsecs is 1/(its parallax in arcseconds). The parallax of Barnard's star is 0.55 arcseconds. This exercise in finding proper motion makes one appreciate the extreme refinement needed to detect such a small angle as this, though in principle the task is the same.

2. As an example of the method of interpolation take line c, which lies between lines B and C of iron. Line c lies to the right of line B at a distance equal to a fraction 0.33 of the distance from B to C. The wavelength difference between B and C is 4.85 nm; therefore line c has a wavelength 0.33×4.85 greater than B, or 397.27 nm.

The Doppler shift $= 397.27 - 396.86 = 0.41$ nm and therefore the velocity is

$$3 \times 10^8 \times 0.41/396.9 = 31 \times 10^4 \text{ m s}^{-1}.$$

(Do not expect to get exactly the same answer for a particular measurement, as each measurement contains its own uncertainty.)

The average velocity from five separate lines (measured by one observer) was $32 \pm 3 \times 10^4$ m s^{-1}. The correct result is 30.4×10^4 m s^{-1}. The shift is towards longer wavelength, i.e., recession.

The star is not a member of the Galaxy; it belongs to the Large Magellanic Cloud, a nearby external galaxy, and its apparent high velocity is largely due to the rotation of our own part of the Galaxy around the galactic centre with a velocity of 22×10^4 m s^{-1}. The radial velocity of a local star with a typical velocity of 3×10^4 m s^{-1} would not be detectable on our photograph. Such measurements require a precise measuring machine which records positions to 0.001 mm on a spectrogram.

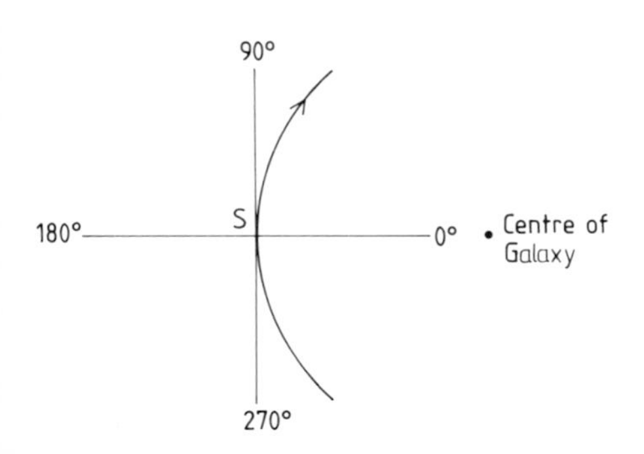

Figure 5.2 Rotation of the Galaxy in the direction of galactic longitude 90°. The Sun is at S. Longitude is measured in the plane of the Galaxy from the direction of the galactic centre which is at longitude 0°.

The Rotation of the Galaxy. When the observed proper motions and radial velocities of large samples of stars in the Galaxy are analysed, certain systematic trends are found. While in the immediate neighbourhood of the Sun, the stars, including the Sun, may be considered as moving at random, on a larger scale in the plane of the Galaxy's disk the pattern of motions varies with galactic longitude and with distance. By combining observations of proper motions and radial velocities a residual motion through space is found for the stars in the solar neighbourhood. The motion is one of rotation around a distant point (identified as the centre of the Galaxy) at the minute rate of 0.0053 arcsec per year. It is also found that on average the stars closer to the galactic centre than the Sun travel faster around the galactic centre than those of the solar neighbourhood, and that those farther out travel more slowly. The rotation of the Galaxy is in the direction of galactic longitude 90° (figure 5.2). A confirmation of this is that the motion of the Sun as it is carried along by galactic rotation shows up as an additional Doppler shift in the spectra of external galaxies; the shift is negative (velocity of approach) in the direction of longitude 90° and positive (velocity of recession) in the opposite direction of longitude 270°. The star in Exercise 2 is an example of this effect; the Large Magellanic Cloud to which it belongs is situated at galactic longitude 280° (and low galactic latitude, −32°). The greater part of the apparent radial velocity of this star is due to galactic rotation.

You will find it easy to calculate that the rotation rate of 0.0053 arcsec per year deduced from observed systematic motions of the stars is equivalent to a period of rotation (one complete circuit of 360°) of 245 million years, usually rounded to 250 million years. The distance to the centre of the Galaxy is 8.5 kpc. Converting this distance to metres (R) and the period of rotation to seconds (P) gives the velocity of rotation of the Sun's neighbourhood around the galactic centre ($2\pi R/P$) in metres per second. The velocity is 22×10^4 m s^{-1}. (See Appendix 1 for the necessary conversions for these calculations.)

The observed rotation of the Galaxy allows the mass of the Galaxy to be calculated by applying Kepler's third law, which states that

$$P^2 = R^3/M$$

where P is the period in years of a body of relatively small mass in orbit around a central body of mass M solar masses at a distance of R astronomical units from the central body. This law, originally formulated in the solar system (§2), can be applied to the Galaxy if most of the mass of the Galaxy is interior to the Sun, in other words, if the Sun is considered as virtually outside the Galaxy. Substituting in Kepler's third law gives 9×10^{10} solar masses for the mass of the Galaxy, rounded to 10^{11} solar masses. This mass is a minimum value, because it does not include those parts of the Galaxy which are further from the centre than the Sun. Though it is relatively easy to record and interpret motions of those parts of the Galaxy which are closer to the galactic centre than the Sun, it is difficult to do the same for positions beyond the Sun. The generally adopted total mass, calculated by a variety of methods, is 2×10^{11}. There is some evidence that the Galaxy may be considerably more massive than this figure.

6 Open Star Clusters

Star clusters are groups of stars which are close together in space. There are two kinds of clusters—open or loose clusters and globular clusters which are tight spherical groups containing enormous numbers of stars. Globular star clusters belong to the halo of the Galaxy. Open clusters, with ages of up to 10^9 years belong to the disk of the Galaxy, and the very youngest among them belong to the spiral arms where star formation still goes on. Two examples of open star clusters are shown in the photographs. Plate 6.1 shows the well known group of the Pleiades, which is visible to the naked eye. The brightest members of the Pleiades are accompanied by bright nebulosities; these are reflection nebulae—dust clouds near the stars which reflect their light. Plates 6.2 and 6.3 show a more distant cluster, NGC 2477 photographed in V (visual or yellow) and B (blue) wavebands.

When a cloud of pre-stellar material condenses it forms fragments which contract under their own gravity until the interiors are hot enough for nuclear reactions to take place. From that moment the objects are real stars. Large masses collapse rapidly—in about 10^5 years; smaller ones less rapidly—in about 10^7 years in the case of a star like the Sun. The energy of a star comes from the inner core of about 20% of the star's mass, where hydrogen is transformed into helium with the release of energy which is radiated upwards to the surface of the star. This is the first phase of a star's life. Theory predicts and observation confirms that the luminosity of a star (that is the rate at which it radiates energy) increases with its mass. The luminosity and surface temperature also go hand in hand: the higher the temperature the greater the rate of radiation per unit area.

A graph showing luminosity against surface temperature for stars in this first phase of their existence is a very smooth curve. It is usually plotted in the form of absolute magnitude M_V against colour $(B-V)$ which is also a measure of its surface temperature, with the most luminous stars (lowest magnitudes) at the top and the colour increasing (redder, cooler stars) to the right. This curve, called the *standard main sequence*, has been built up from observations of stars in nearby clusters whose distances are known (figure 6.1). All stars, from the time of their birth until the core has used up its hydrogen supply, lie on this curve. The luminosities of stars on the main sequence range from a million solar luminosities to a tenth of a solar luminosity.

The supply of hydrogen to fuel a star is proportional to its mass; however, the rate at which a star expends its energy per unit mass increases very greatly with its mass. The time taken to use up the fuel is proportional to the mass divided by the luminosity. A star at the top of the main sequence goes through its main sequence phase in only 1 or 2 million years. A star like the Sun lasts 10 billion years on the main sequence. The least massive stars expend their energy so slowly that their main sequence lifetime is 50 billion years or more, longer than the present age of the Galaxy itself.

When a star comes to the end of its main sequence lifetime it goes through another stage. It expands and becomes cooler though remaining luminous, while the helium in its interior is transformed into other elements and releases more nuclear energy. A star in this phase is a *red giant*.

The shape of the main sequence is the same for all star clusters of whatever age, with only minor variations. This fact provides a very important means of finding the otherwise unknown distance of a cluster. A plot of apparent magnitude against colour for stars in a cluster is called the *colour–magnitude diagram*. The part of this diagram occupied by main sequence stars has the same shape as the standard main sequence. However, the magnitudes on the cluster's diagram, being apparent magnitudes, are all larger than those at the corresponding positions on the standard curve, which are absolute magnitudes. By definition (Appendix 2), the difference between apparent and absolute magnitude (the distance modulus) is

$$5 \log (r/10)$$

where r is the distance in parsecs. In the case of a star cluster, the distance modulus is the shift in the vertical axis needed to bring its colour–magnitude diagram into coincidence with the standard main sequence. (Interstellar extinction affects the magnitudes and colours of stars; this has first to be allowed for. Details of how to do this are omitted here.) The cluster may, on account of age, have lost its most luminous stars; however, stars which belong to the lower part of the main sequence are always present however old the cluster may be.

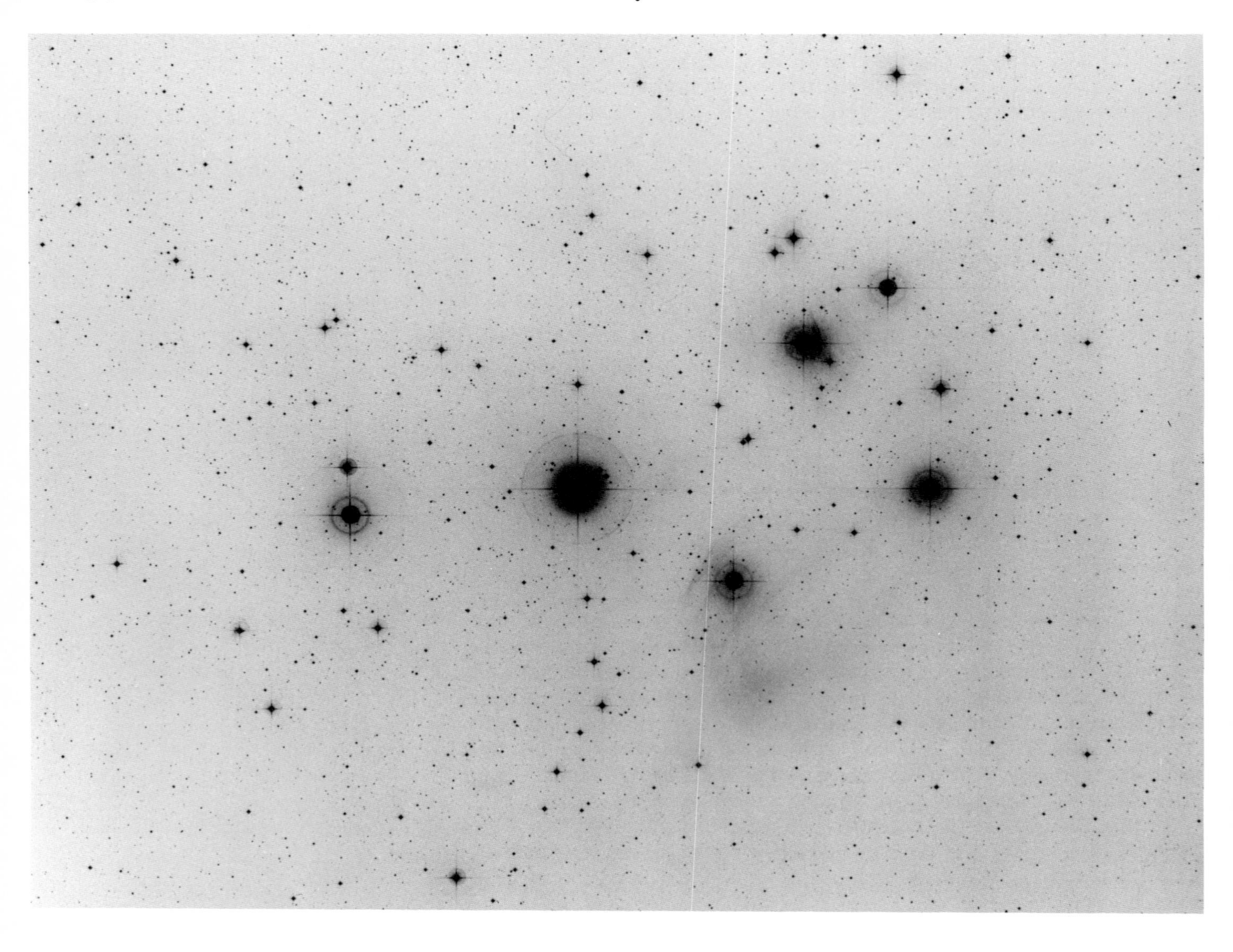

Plate 6.1 The star cluster of the Pleiades in yellow (V) light (© Royal Observatory, Edinburgh).

A second important piece of information which comes out of the comparison of the cluster and standard main sequences is the age of the cluster. The lifetimes on the main sequence of stars of various luminosities are known from theory. The most luminous star which still remains on the main sequence thus indicates the age of the cluster, because all the more luminous ones have already evolved. In some cases the stars which are near the top of the diagram lie slightly off the main sequence; these are at the end of their main sequence phase and are about to leave or 'turn off' the main sequence. The turn-off point is thus an age indicator.

Good photometry is required for building up colour–magnitude diagrams; the photometric accuracy which we were able to achieve in §4 is not sufficient to give a meaningful colour–magnitude diagram, and in the exercises which follow we make use instead of published photoelectric photometry.

Exercise 1. Draw the colour–magnitude diagram of the Pleiades from the data in table 6.1 in the form of absolute magnitude (M_V) against ($B-V$). The distance of the Pleiades is 127 pc. Demonstrate that its colour–magnitude diagram coincides with the standard main sequence.

Table 6.1

V	B
2.87	2.78
3.64	3.56
3.71	3.60
3.88	3.81
4.18	4.12
4.31	4.20
5.09	5.01
5.45	5.38
5.76	5.72
6.29	6.31
6.82	6.84
6.99	7.02
7.35	7.45
7.66	7.87
7.85	8.05
8.12	8.34
8.27	8.63
8.37	8.67
8.69	9.15
9.25	9.80
9.45	9.97
9.88	10.42
10.13	10.75
10.48	11.12
10.83	11.63

(All the stars are on the field of plate 6.1. The first six are the naked eye stars.)

Suggestions. Subtract the first column from the second to get ($B-V$) and plot V against ($B-V$) following figure 6.1 and using if possible the same scale. You will notice that the scale is such that an interval of 1 magnitude on the vertical scale corresponds to 0.2 magnitude on the horizontal scale. To convert absolute magnitudes into apparent magnitudes, subtract

5 log (12.7). It is not necessary to subtract this from each individual V; it is easier to shift the vertical scale on the V, $(B-V)$ diagram after it is plotted.

THE AGE AND DISTANCE OF A CLUSTER

Plates 6.2 and 6.3 show a typical example of an open star cluster photographed in the conventional wavebands V and B. The cluster members abound in the central regions, but most of the stars in the outer parts of the photographs belong to the general field of the Galaxy. If you look closely at the two photographs and compare the images of bright stars (i.e. the large images) individually you will notice that there is a greater proportion of stars in the upper range of brightness on the V photograph than on the B photograph. The reason for this is that there are several red giant stars in the cluster which are considerably brighter in the V waveband than in the B. Those stars are clearly seen on the colour–magnitude diagram of the cluster, plotted in figure 6.2. The bunch of points in the middle of the diagram, for which $(B-V)\simeq 1$ indicate stars which are about 1 magnitude brighter in V than in B. As in Exercise 1 we use photoelectric photometry for this cluster, and, to simplify proceedings, the individual magnitudes are not listed but are plotted as a colour–magnitude diagram in figure 6.2 for the brightest members. Magnitudes (V) and colours $(B-V)$ have already been corrected for interstellar extinction.

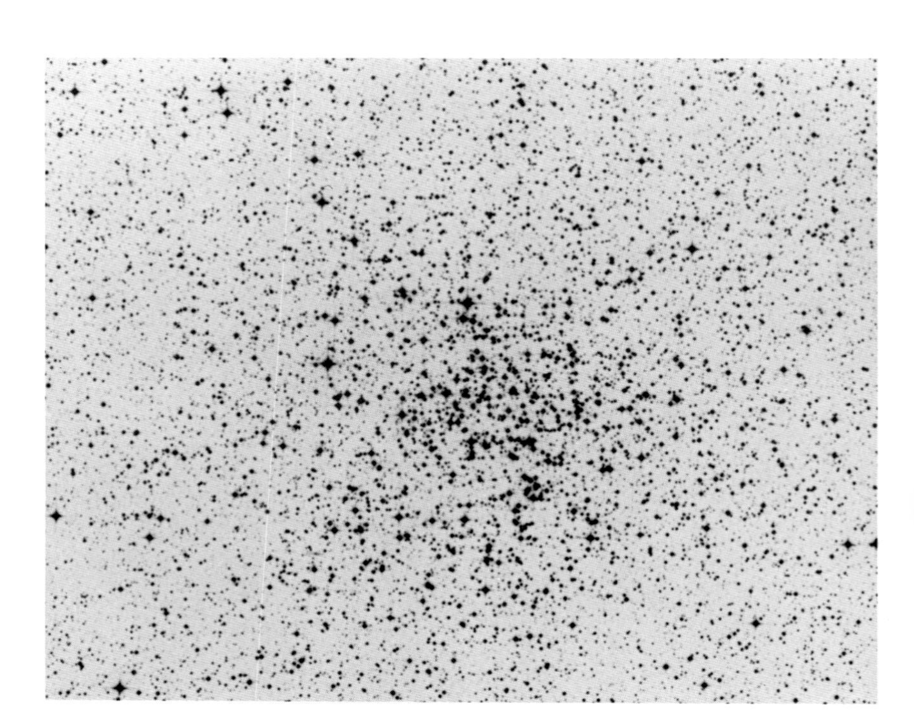

Plate 6.2 The open star cluster NGC 2477 in yellow (V) light (© Royal Observatory, Edinburgh).

It has already been explained that luminous stars complete their lives on the main sequence sooner than less luminous ones. Table 6.2 gives the maximum ages of main sequence stars as a function of their absolute magnitudes.

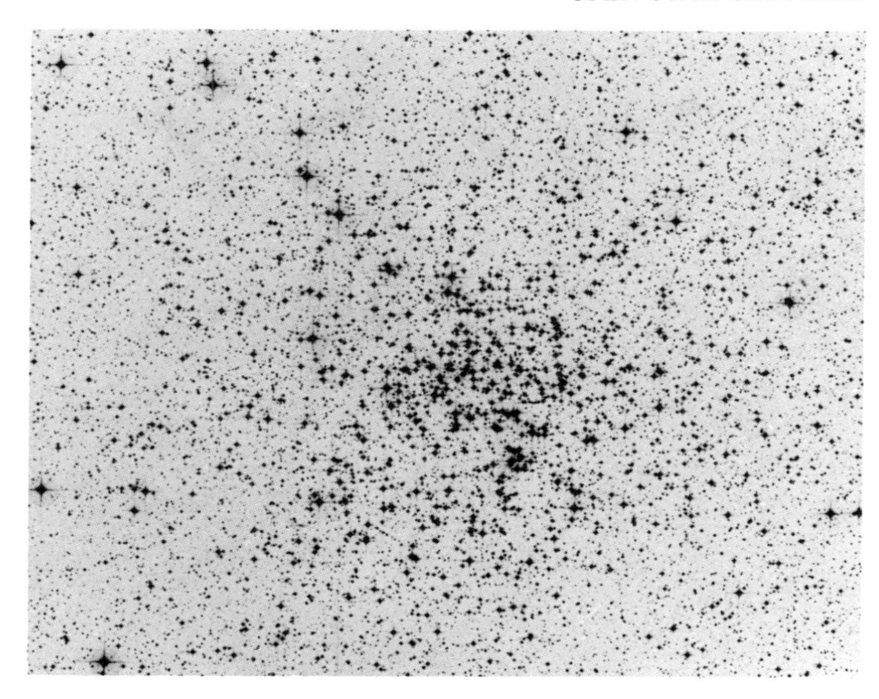

Plate 6.3 The cluster NGC 2477 in blue (B) light (© Royal Observatory, Edinburgh).

Table 6.2

Absolute magnitude M_V	Log (age in years)
−7	6
−4	7
−1	8
+2	9

Exercise 2. Figure 6.2 shows the colour–magnitude diagram of the brightest stars in the cluster NGC 2477. Find the distance of the cluster by comparing the diagram with the main sequence, and find also its age by making use of the information in table 6.2.

Suggestions. Trace the main sequence from figure 6.1 on transparent paper. Draw in also the axes and mark the magnitudes M_V and the colours $(B-V)$. Superimpose this on the colour–magnitude diagram with the $(B-V)$ axis exactly lined up. Move the graph vertically until the main sequence fits as well as possible on the points of the diagram. (Ignore the red giant stars, that is, those points which lie to the right side of the diagram.) Read off to the nearest 0.1 magnitude the point on the vertical axis of the diagram which corresponds to absolute magnitude 0 on the standard main sequence. This is the distance modulus. Convert this to parsecs using the standard formula given earlier.

Note also on the main sequence the point near the top of the colour–magnitude diagram where it begins to diverge from the standard main sequence, and read off the absolute magnitude at this point. The relation between the turn-off point and age is given on table 6.2. You may need to interpolate between two values. Do this in logarithms, i.e., plot a graph of log (age) against turn-off magnitude; read off the log (age) corresponding to the turn-off magnitude and convert to age in years.

SOLUTIONS AND DISCUSSION

1. The colour–magnitude diagram of the Pleiades obviously coincides with the standard main sequence when shifted by the distance modulus 5.5 mag. The Pleiades cluster has lost its most luminous stars because of its age (5×10^7 years). We have used only 25 stars in the

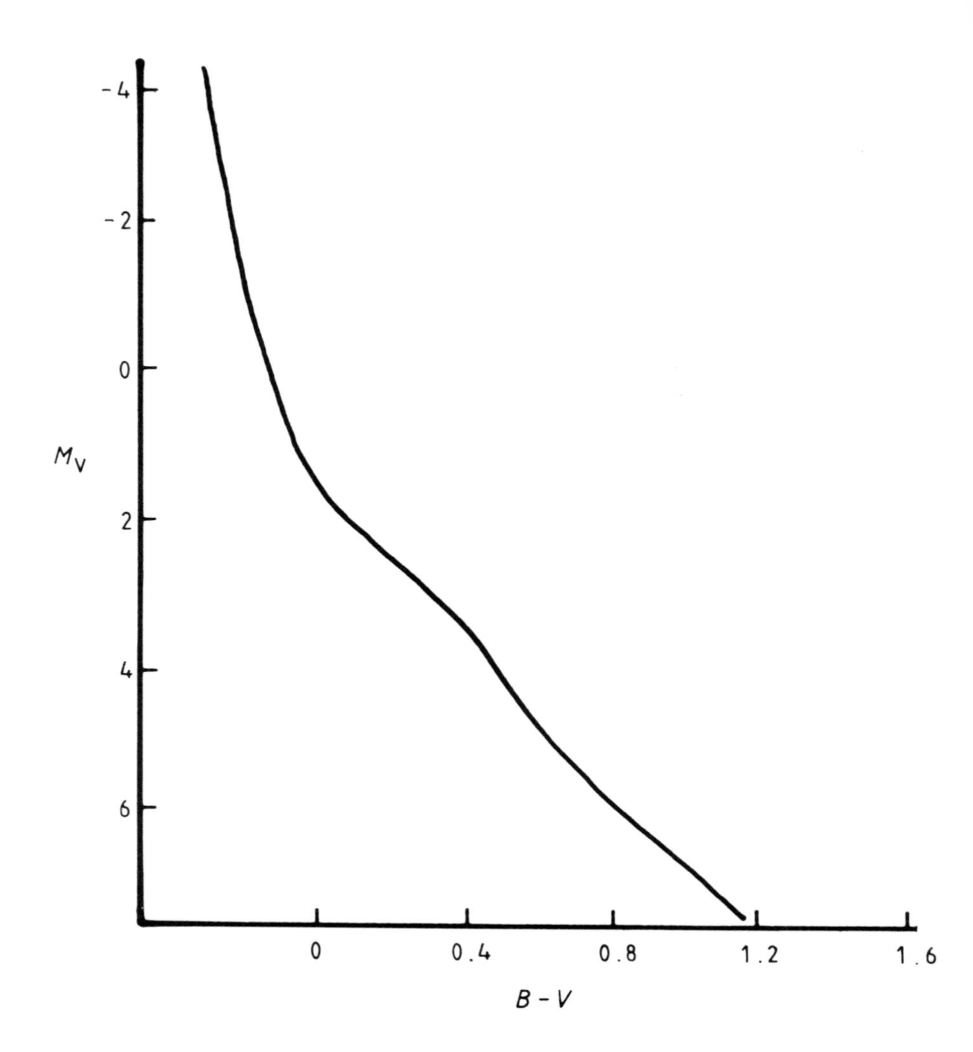

Figure 6.1 The standard main sequence.

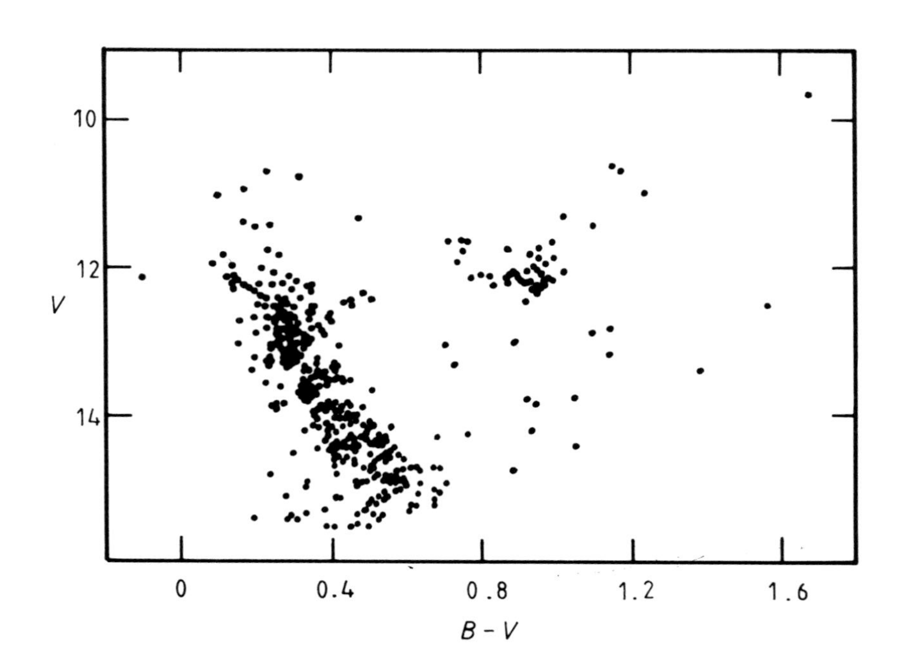

Figure 6.2 The colour–magnitude diagram of the cluster NGC 2477 at (7h 52m, −38.5°).

exercise; in fact the observed colour–magnitude diagram of the Pleiades extends several magnitudes fainter than these and contains about 250 members with measured photoelectric magnitudes.

2. The distance modulus of the cluster NGC 2477 is 10.6 and its distance is 1300 pc (antilog $((10.6+5)/5)=1318$). The age, from the turn-off point, is 9×10^8 years. These figures come from a more detailed colour–magnitude diagram than the one which we have used, which reaches fainter stars, and to which a theoretical curve was fitted. You can assess the accuracy of the distance and age which you have found from the accuracy with which you have been able to fix the distance modulus.

Colour–magnitude diagrams of star clusters have to be corrected for interstellar extinction, that is, for the dimming of starlight by dust in the space between the stars, mentioned in the Discussion in §4. Dust is concentrated in the galactic plane, and, since open star clusters lie in that plane, all but the nearest of them are affected by extinction. In the early days of galactic research it was noticed that cluster distances, estimated from the apparent magnitudes of member stars, were systematically larger than distances estimated from the apparent cluster diameters, assuming that all open clusters have more or less the same linear dimensions. The apparent diameter of a cluster decreases inversely as the distance, while the apparent brightness of a star decreases inversely as the square of the distance. The observation that cluster stars appeared dimmer than expected was one of the pointers to the existence of obscuring material in the intervening space.

The effect of extinction is to make stars appear dimmer and redder than they really are (§8). Magnitudes (V) and colours ($B-V$) in the colour–magnitude diagram of the cluster NGC 2477 used in the exercise are already corrected for this effect. The correction is 1.7 magnitudes in V, which means that if extinction had not been allowed for, the cluster would have been placed at more than twice its true distance ($10^{(1.7/5)}=2.2$). The colour–magnitude diagram of the Pleiades, a nearby cluster, does not suffer from interstellar extinction.

The method of finding the distance of a cluster by matching the main sequence is a reliable one, because the use of a large number of stars smoothes out the photometric errors in the individual points. Investigations of the distances of open star clusters have played an important part in tracing the spiral arms of the Galaxy. The galactic coordinates of NGC 2477 (354°, −6°) indicate that it is close to the galactic plane and not far from the direction to the galactic centre. At a distance of 1.3 kpc from the Sun it belongs to an inner spiral arm of the Galaxy, adjacent to the one in which we are situated.

Modern research on colour–magnitude diagrams is concerned not only with age and distance but also with small variations in the evolution of stars in clusters of different chemical composition.

7 Globular Star Clusters

Globular clusters are the most conspicuous components of the halo of the Galaxy. They are aggregations of many thousands of stars swarmed together in regular spherical shapes. About 150 are known, though many more are undoubtedly concealed in remote parts of the Galaxy, behind the absorbing dense regions of the galactic centre. The two brightest globular clusters are Omega Centauri (NGC 5139) and 47 Tucanae (NGC 104), both situated in the southern celestial hemisphere in the constellations respectively of the Centaur and the Toucan. These, and a few less conspicuous ones such as M13 (NGC 6205) in the northern constellation of Hercules are visible as fuzzy blobs to the naked eye. The centre of the globular cluster system identifies the centre of the Galaxy as a whole.

Globular star clusters contain variable stars (stars with periodically varying brightness) of a type called RR Lyrae (called after the star of that name which has given its name to all of the same genre). RR Lyrae stars have been the basic means by which globular cluster distances have been determined. Modern instruments can also record the faint unevolved stars in many globular clusters giving an independent confirmation of their distances by the method used for open star clusters (§6).

The photograph (plate 7.1) shows the globular cluster NGC 362, at galactic coordinates (301°, −47°). The cluster is at a distance of 9.3 kpc from the Sun. The same cluster, and the large 47 Tucanae, appear on a smaller scale on plates 11.2 and 11.4.

From their regular and compact shape it is surmised that globular clusters are stable, that is, that their member stars hold together as a body. The stars are not at rest within the cluster; like all stars everywhere, they move about among each other. Globular clusters are among the oldest objects in the Galaxy. They have therefore held together for over 10^{10} years. In that time the stars have moved around a lot within the cluster; they have come close to each other and been deflected by each other's gravitational influence many times. The result is that their kinetic energies (energies of motion) are divided out between them in a certain statistical manner, some with more, some with less than the average kinetic energy. In a system like a globular cluster which holds its members by gravitation, the total kinetic energy has to be less than the potential energy of gravitation, otherwise the stars would by now have flown apart. It is possible to calculate the mass of a cluster by making use of this principle. The method involves finding the cluster's kinetic energy from Doppler shifts in the spectra of its stars, and calculating the potential energy from the distribution of stars within it. The method is demonstrated in §12 with an example from a cluster of galaxies.

There is an alternative way of estimating the mass of a globular cluster which forms the basis of the exercises which follow. A globular cluster experiences the gravitational influence of the Galaxy as a whole. It moves in an orbit around the centre of the Galaxy, just as a planet moves around the Sun. However, the cluster has a significant size, so that the points within it are at different distances from the centre of the Galaxy. According to the law of gravitation the force of gravitational attraction decreases as the inverse square of the distance from the massive body; therefore the stars in

Facing page:

Plate 7.1 The globular cluster NGC 362 (© Royal Observatory, Edinburgh).

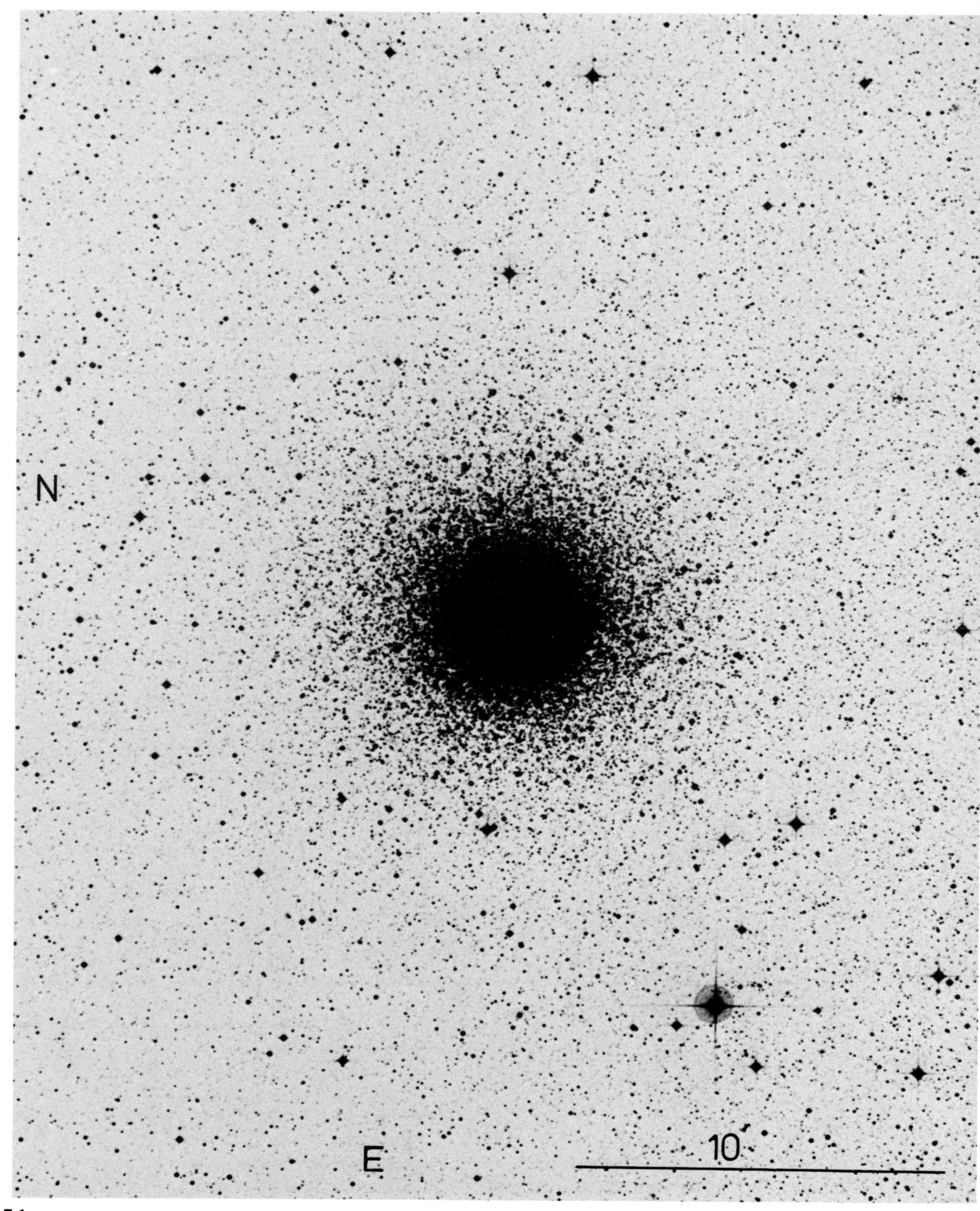

Plate 7.1

the cluster which are nearest to the centre of the Galaxy are the most strongly attracted and those which are most distant are the least strongly attracted. The outermost stars in the cluster would become detached as a result of this difference in gravitational force if they were not held back by the cluster's own gravity. The edge of the cluster is where the two effects are balanced, and is called the 'tidal' limit by analogy with the tides on the Earth's oceans which are also caused by differential gravitational effects (of the Moon on the waters).

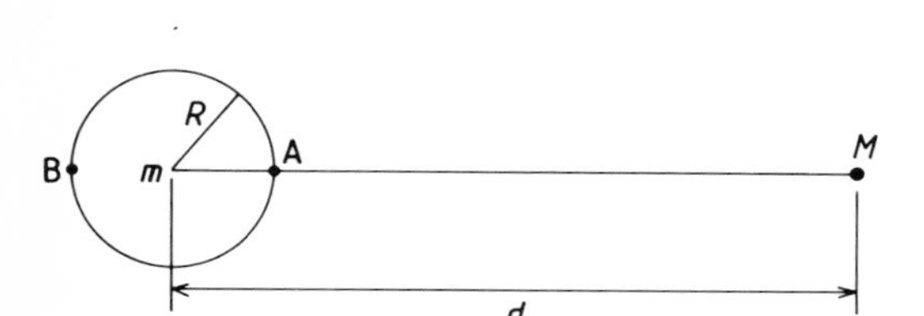

Figure 7.1 Diagram of a globular star cluster shown as a sphere of radius R and its position relative to the centre of the Galaxy, shown as a point on the right at a distance d from the cluster. M and m are the masses respectively of the Galaxy and the cluster.

Figure 7.1 is a diagram showing the cluster and the centre of the Galaxy which as far as its gravitational influence is concerned acts as if all the Galaxy's mass were situated there at one point. The limiting radius of the cluster is R, the distance of the cluster from the centre of the Galaxy is d, and the masses of the cluster and the Galaxy are respectively m and M. A unit mass at A on the outermost point of the cluster is attracted by the Galaxy with a force

$$GM/(d-R)^2.$$

A unit mass at B on the opposite point of the cluster is attracted by the Galaxy with a force

$$GM/(d+R)^2.$$

The difference between these two forces is a force separating the two masses from each other. That difference is

$$GM/(d-r)^2 - GM/(d+R)^2$$
$$= GM(4dR)/[(d-R)^2(d+R)^2].$$

The radius of the cluster R is very small compared with the cluster's distance from the galactic centre, so R may be left out in the denominator. The result then becomes $4GMR/d^3$. (The formula is not exactly correct because it does not take into account that the cluster is in orbit around the centre of the Galaxy. A better formula replaces 4 by 3.5.)

This difference is a force which drives the two masses apart. It is balanced by the gravitational pull of the cluster itself, which attracts each of the unit masses towards its centre with a force Gm/R^2. At the tidal limit the two forces exactly balance, so

$$3.5GMR/d^3 = Gm/R^2$$

or

$$m/M = 3.5(R/d)^3.$$

In this formula the mass of a cluster comes out in terms of the mass of the Galaxy and of the ratio of the radius of the cluster to its distance from the centre of the Galaxy.

The important measurement to be made is the radius of the cluster. There are many stars in the surroundings of the cluster which means that the position of its edge is far from obvious. The numbers of stars per unit area at different distances from the main body of the cluster have to be counted, until the position is reached when the numbers level off to a constant value.

Exercise 1. Find the tidal radius of the globular cluster NGC 362. Give the result in parsecs, given that the distance of the cluster from the Sun is 9.3 kpc.

Suggestions. First decide on a suitable unit of area for counting purposes: if this is too small there will be too few stars; if too large, the measurement of radius will be too coarse. Try squares of 5×5 or 10×10 mm. Decide also on the limit of faintness to which you intend to count. The fainter you go, the higher the counts and the better the result; be sure, however, that you are consistent in counting everywhere always to the same limit of faintness.

There is no point in counting stars in the crowded inner regions since we are interested only in finding the edge, so concentrate on the outer regions. One way to go about the counting is to cut out of a sheet of paper an opening in the form of a long narrow strip of width 5 or 10 mm, depending on the size of square you intend to use (figure 7.2). Lay this across a diameter of the cluster as centrally as you can. Count the number of stars which you see in the gap in each unit interval along the strip; it is quite easy to judge each square by eye, but you may want to cover up each bit as soon as you have counted it, to avoid confusion. Begin well away from the cluster at one side and number the squares as you go along, noting down the number of stars in each square. When it is obvious that the star counts have gone up significantly, begin counting the opposite side, keeping the sequence of numbering of the squares and continuing until you are well away from the cluster on that side also.

Repeat the process across another diagonal—and across as many others as you have time and patience for.

For each set, plot the star numbers against the running number of the square, and decide where the numbers definitely level off to a constant value on each side. You may not get the same result from all the diagonals. You may also be troubled by the fact that background counts are not the same on all sides—this is partly inherent in the statistics and partly (in this particular case) because some of these stars belong to the Small Magellanic Cloud, a galaxy which happens to be in the same part of the sky. The expected background count may be checked by counting larger sample areas at the extreme edges of the photograph, well away from the cluster.

Look at all your results and take the mean radius. Remember that you have made your measurements in your chosen units which must be converted to angular measure using the scale on the photograph. This in turn must be converted into radians (Appendix 1). To get the radius in parsecs multiply the radius in radians by the distance to the cluster in parsecs.

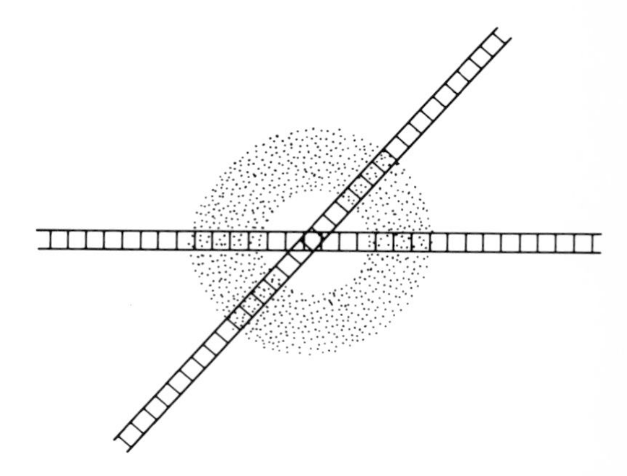

Figure 7.2 A method of counting stars in a globular cluster.

Exercise 2. Using the tidal model calculate the mass of the cluster in solar masses. The distance of the Sun from the galactic centre is 8.5 kpc and the mass of the Galaxy is 2×10^{11} solar masses. The galactic coordinates of the cluster and its distance from the Sun have been given already.

To get the mass is a matter of substitution in the formula for m/M given earlier. First, however, you need to know d, the distance of the cluster from the galactic centre. This may be done either graphically or by trigonometry. You can get the angular separation between the two points on the celestial sphere from a globe (in a procedure similar to that used in §3). Alternatively, you may calculate it from a very simple formula for the angular distance θ of a point at galactic coordinates (l, b) from the galactic centre (coordinates $(0, 0)$):

$$\cos \theta = \cos l \cos b.$$

The Sun, the cluster and the galactic centre form a triangle which has an angle θ at the Sun (figure 7.3). The distance d required in the calculation is found by solving this triangle either graphically or by trigonometry.

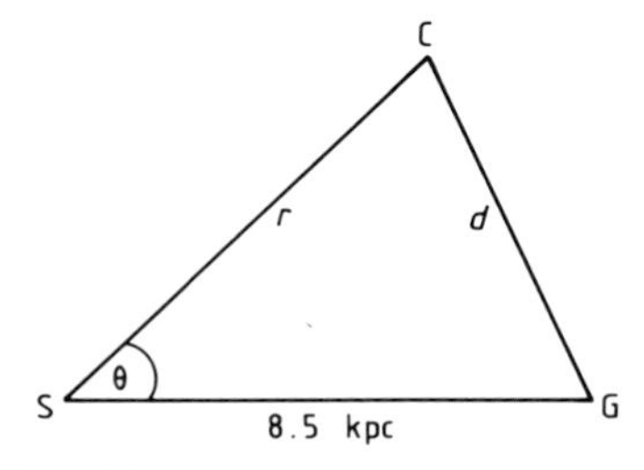

Figure 7.3 Triangle showing the relative positions of the Sun (S), the cluster (C) and the galactic centre (G) from which the distance d of the cluster to the galactic centre is found.

Suggestions. The cluster, at galactic longitude 301°, is 59° from the galactic centre. Its latitude is $-47°$. To get the angular distance θ between the cluster and the galactic centre from a globe mark a point on the equator to represent the galactic centre. Move 59° along the equator and 47° upwards or downwards (direction does not matter here) to represent the position of the cluster. Join the two points and measure their angular separation. Draw to scale (1 cm = 1 kpc) on graph paper a line joining the Sun and the galactic centre, and another line at an angle θ from the Sun, marking off the position of the cluster. Join the cluster to the galactic centre, completing the triangle (figure 7.3). Measure this distance with a ruler and convert back to parsecs. Substitute this, and also the radius of the cluster in parsecs from the previous exercise, in the formula to get (m/M), the mass of the cluster as a fraction of the mass of the Galaxy. Multiply by the mass of the Galaxy in solar masses to get the mass of the cluster in the same units.

Exercise 3. Estimate the average star density in the cluster NGC 362 and the average distance between stars, assuming that the mass of an average star in the cluster is approximately one solar mass.

SOLUTIONS AND DISCUSSION

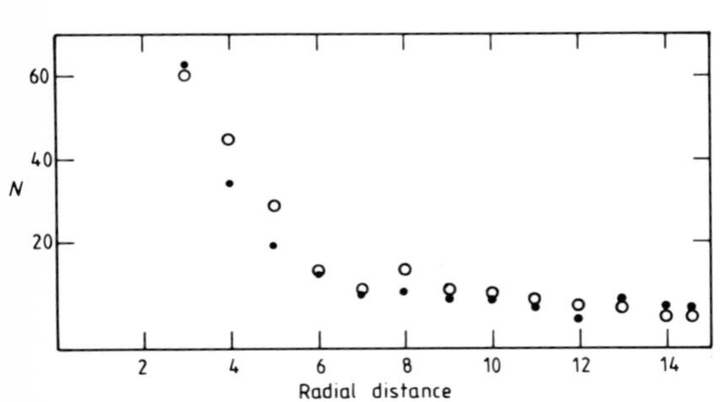

Figure 7.4 An example of star counts in two radial directions from the cluster. Radial distance is in arbitrary units. The counts level off to a uniform value beyond the limit of the cluster.

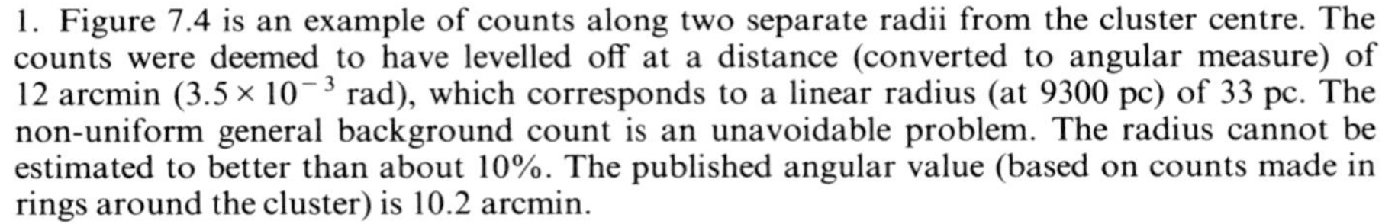

1. Figure 7.4 is an example of counts along two separate radii from the cluster centre. The counts were deemed to have levelled off at a distance (converted to angular measure) of 12 arcmin (3.5×10^{-3} rad), which corresponds to a linear radius (at 9300 pc) of 33 pc. The non-uniform general background count is an unavoidable problem. The radius cannot be estimated to better than about 10%. The published angular value (based on counts made in rings around the cluster) is 10.2 arcmin.

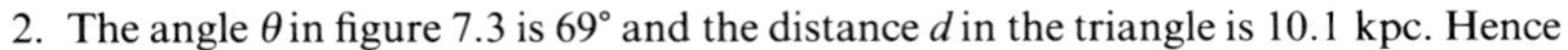

2. The angle θ in figure 7.3 is 69° and the distance d in the triangle is 10.1 kpc. Hence

$$R/d = 32.7 \times 10^{-4}$$

substituting in the formula gives

$$m/M = 2.4 \times 10^4 \text{ solar masses.}$$

We should remember that the mass is proportional to the cube of (R/d); hence any error in R/d becomes much larger in the mass—in fact, three times larger; if we estimate that there is an error of 10% in R, the error in the mass is 30%. In view of this uncertainty, the result may be given as between 1 and 4×10^4 solar masses.

The mass of NGC 362, obtained by the more refined method mentioned in the text, is 1.8×10^5 solar masses. There are possible explanations for the discrepancy: the counts may not have quite reached the tidal limit, and the model of the cluster with its tidal cut-off is undoubtedly too simple. However, it is encouraging that the results agree within an order of magnitude.

3. If the stars are uniformly distributed inside the spherical cluster, the density is the mass divided by the volume. The volume is $\frac{4}{3}\pi R^3 = 1.5 \times 10^5$ pc^3; hence the density is 0.16 solar masses per pc^3. If each star has a mass of 1 solar mass on average, there are 0.16 stars per pc^3. The volume of space occupied by each star on average is the reciprocal of this (1/0.16) or 6 pc^3. If we imagine this volume as a cube, the side of the cube is the cube root of 6 = 1.8 pc, which is also the distance of each star from its neighbours. This distance is about half the average separation of stars in the solar neighbourhood.

It is known that the density at the centre of a globular cluster is higher than the average for the cluster as a whole—perhaps as high as 100 times the average—which would make the separation of the stars in the centre about five times less (cube root of 100 approximately), or about 0.4 pc. This is still quite a large distance and demonstrates that stars within globular clusters are not as densely packed as they seem to be from most photographs. Photographs taken under the best conditions of 'seeing' (with very small star images) reveal gaps between the stars in the very core of globular clusters (plate 7.2).

Plate 7.2 The central region of the globular cluster NGC 288 at (0h 53m, −26.5°) in which gaps are seen between the stars (© Royal Observatory, Edinburgh).

8 Interstellar Extinction

The space between the stars in the disk of the Galaxy is not empty. It contains gas and obscuring material. As already discussed (§4), one result of interstellar obscuration is that early estimates of distances based on star counts in the plane of the Galaxy were systematically too large.

Photography of star fields in the Milky Way, begun about 1890, first revealed what are obviously opaque clouds in front of the background multitudes of stars. Dark clouds are very numerous in the southern Milky Way. The largest example is the 'Coalsack', a region of about 40 square degrees where the paucity of stars is evident to the naked eye. An example of obscuration in the northern Milky Way which is also visible to the naked eye is the so-called Rift in the constellation Cygnus. It is an extended dark cloud or chain of clouds which lies along the galactic belt dividing it into two for about 30 degrees. The rift which extends from approximate coordinates (18h, 0°) to (20h, +45°) is shown on good star atlases.

The dimming or extinction of light in the plane of the Milky Way varies greatly from place to place but on average is about 1 mag kpc^{-1}. Stars at a distance of 1 kpc appear about 1 mag fainter than they really are; those at 2 kpc appear about 2 mag fainter and so on. A magnitude is equivalent to a factor of approximately 2.5 in brightness; extinction of 1 magnitude means that stars appear 2.5 times fainter than they really are or that their light is reduced to 40% of its unobscured intensity. Light traversing 2 kpc is reduced by a further factor of 2.5 to 0.4×0.4 or 0.16 of its original brightness. The inner parts of the Galaxy, many kiloparsecs away, are dimmed to invisibility (in ordinary light).

The interstellar obscuring material not only dims the light; it also reddens it. The colours of distant stars in the galactic plane are more red than their spectra would predict. It is usual to describe these stars as 'reddened', though this is a misnomer, because their redness is not due to additional red radiation but to a deficiency at shorter wavelengths. The actual fraction of light removed increases steadily with decreasing wavelength. The observed relation between the extinction and wavelength, the extinction or reddening law, shows that the extinction in magnitudes increases practically linearly with the inverse of the wavelength or $1/\lambda$ (λ being the symbol for wavelength). Observations at infrared wavelengths are particularly useful for investigating heavily obscured parts of the Galaxy. Best of all for penetrating the interstellar fog are radio waves.

The reddening law is explained by the scattering of light by dust in the form of solid grains roughly the size of the wavelength of violet light, that is, about 0.3 microns (3×10^{-7} m). The effect of such small grains is to scatter the light, that is, to reflect it in all directions (apart from a small fraction absorbed by the grains). The shorter the wavelength, the more light is scattered away from the direction in which the original light is travelling. Detailed studies of interstellar extinction suggest that the likely constituents of the grains are silicates and graphite.

THE EXTINCTION LAW

The photographs (plates 8.1, 8.2 and 8.3) of the same star field in the plane of the Milky Way show typical patches of obscuring cloud superimposed on a dense background of stars. The photographs were taken in three different wavebands; blue (J) centred on 420 nm, red (R) centred on 640 nm and near-infrared (I) centred on 800 nm.

Stars seen in the direction of a dark cloud consist of two components: the foreground stars, that is, those which are nearer to us than the cloud; and the background stars which are behind the cloud and are seen through it in dimmed form. It is not possible to say simply by looking at a dark cloud which stars are in front and which are behind it; this is so because the distance of a dark cloud is not obvious from its appearance. In practice, however, clouds which have distinct outlines are likely to be relatively close, within a few hundred parsecs. The clouds in the photographs belong to the complex of the Coalsack, mentioned earlier. It is known from observations of individual stars in front of and behind the Coalsack that the obscuring material there is only about 200 pc away.

We assume that all the stars visible in the obscured region are behind the cloud. (We shall satisfy ourselves that this assumption is reasonable from the data given in the Discussion; you may like to accept it for the moment.)

We already know how the density of stars in the general field increases as one counts to fainter magnitudes (§4). The star density in the region of the dark cloud also increases with magnitude at the same rate, but here all magnitudes are fainter than in the surrounding field on account of the intervening cloud. If the rate at which the density increases with magnitude is known, then, conversely, an observed difference in density can be converted into a difference in magnitude. Figure 8.1 shows this schematically. Line a shows the increase in star numbers per unit area with magnitude in the unobscured region; line b shows the increase in the obscured region. The horizontal shift between the two lines represents the extinction caused by the cloud.

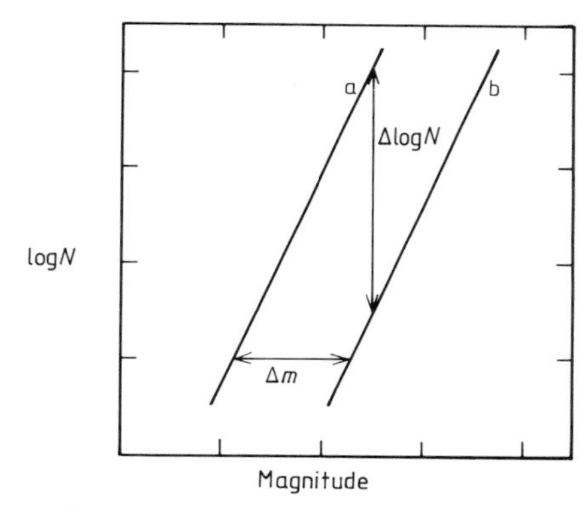

Figure 8.1 Logarithmic increase in the number of stars per unit area of sky with magnitude (*a*) in the general field and (*b*) within a dust cloud. $\Delta \log N$ is the difference in the logarithm of the numbers outside and inside the cloud; Δm is the extinction in the cloud (the symbol Δ means difference).

Exercise 1. Measure the extinction in an interstellar cloud by the method of star counting at three different wavebands and find its dependence on $1/\lambda$, the inverse of wavelength. The rate of increase in log N with magnitude in the general field is 0.4 per magnitude, and is assumed to be the same at all wavelengths. The wavelengths on which the wavebands are centred have already been given.

Suggestions. Choose an area within a cloud of conspicuous obscuration on the blue photograph; the large cloud in the upper part of the photograph is very suitable. The area should be as large as possible to maximise the accuracy of the count, but its shape is irrelevant. Cut out a window of a convenient size and shape. Now decide on the faintness to which you think you can count and choose a limit by the size of the images. The fainter you count the better the accuracy of the counts from the statistical point of view; however it is also essential that you are able to recognise the limit consistently.

Place the window on the cloud and count all the stars to the chosen limit. Move the window to a comparison field away from the cloud, and count again. There are so many stars in the unobscured areas that you may need to subdivide the area in order not to lose count. Repeat this in a number of other comparison areas and take the mean. The more comparison fields you use, the better the accuracy of the mean.

Divide one count by the other to get the ratio of field to cloud numbers, and take the log of this ratio. (Note that the area of the window does not enter into the calculations.) From the known rate of increase in log N calculate the magnitude difference corresponding to the log of the ratio of field to cloud numbers which you have counted. (Remember that the difference in the logarithm of two numbers is the same as the logarithm of their ratio.) This magnitude difference is the extinction.

Page 54:

Plate 8.1 Photograph of a dusty region in the plane of the Galaxy in blue (B) light. A distinct dust cloud is seen in the upper part of the field (© Royal Observatory, Edinburgh).

Page 55:

Plate 8.2 The same field as plate 8.1 in red (R) light (© Royal Observatory, Edinburgh).

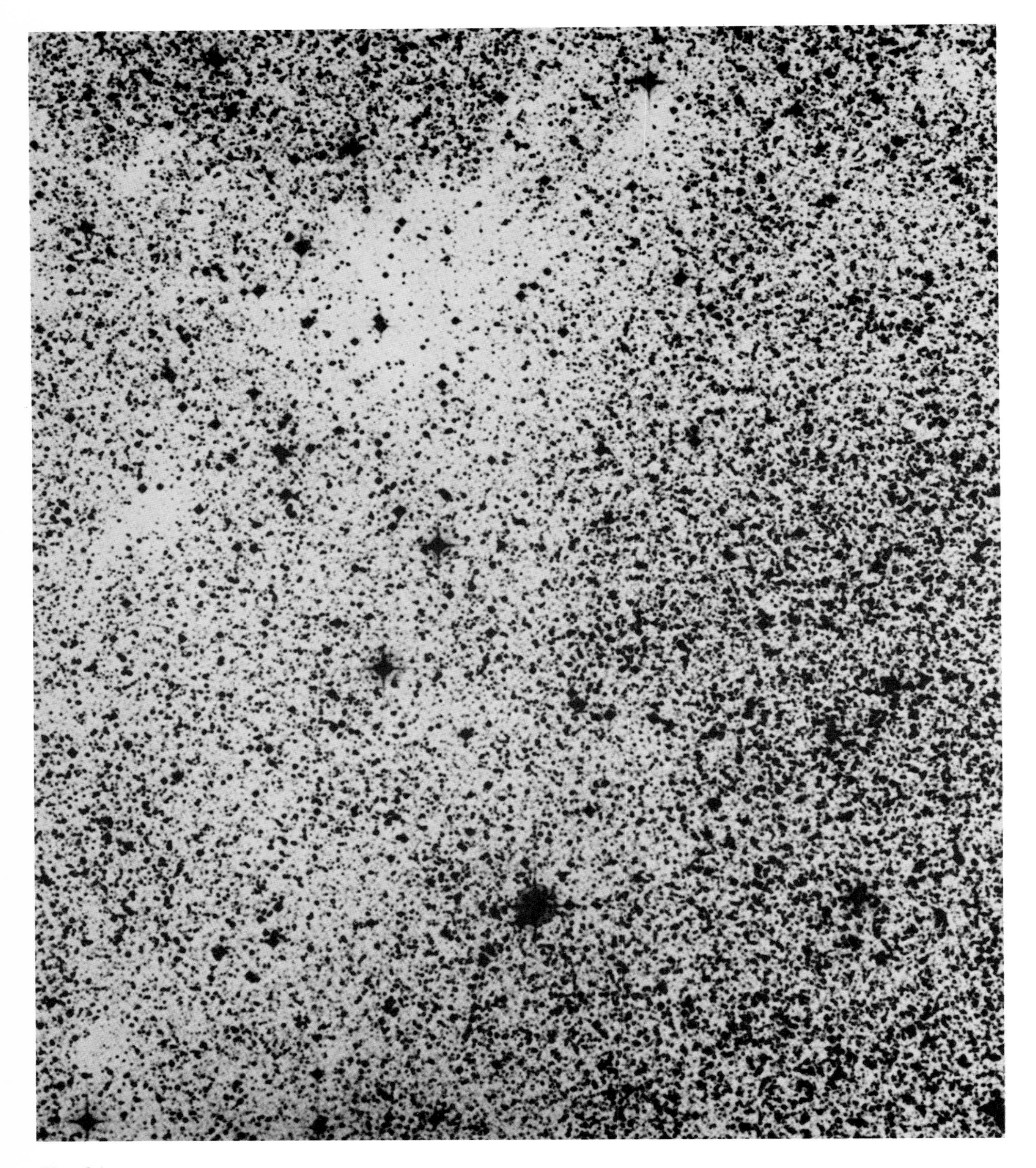

Plate 8.1

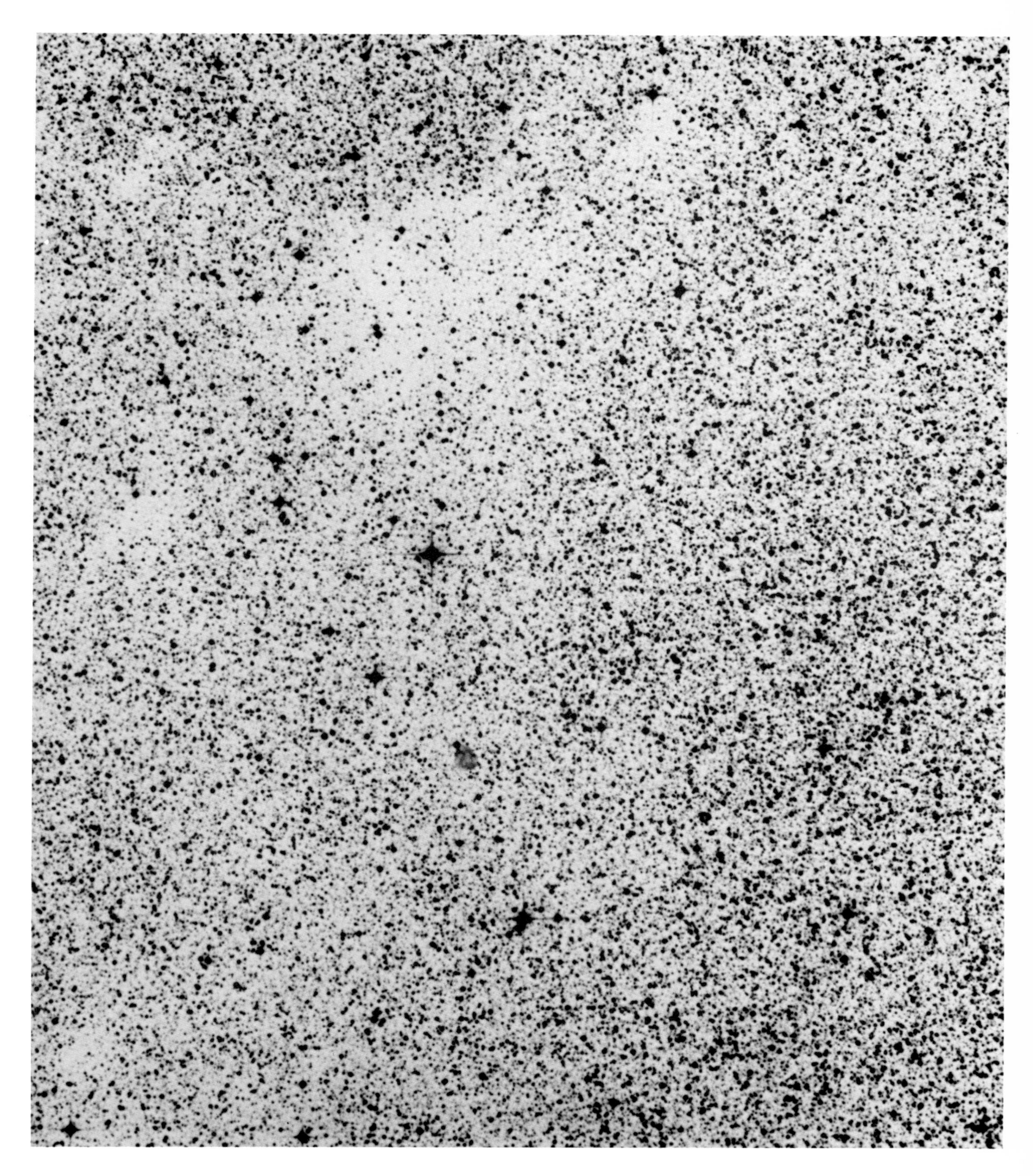

Plate 8.2

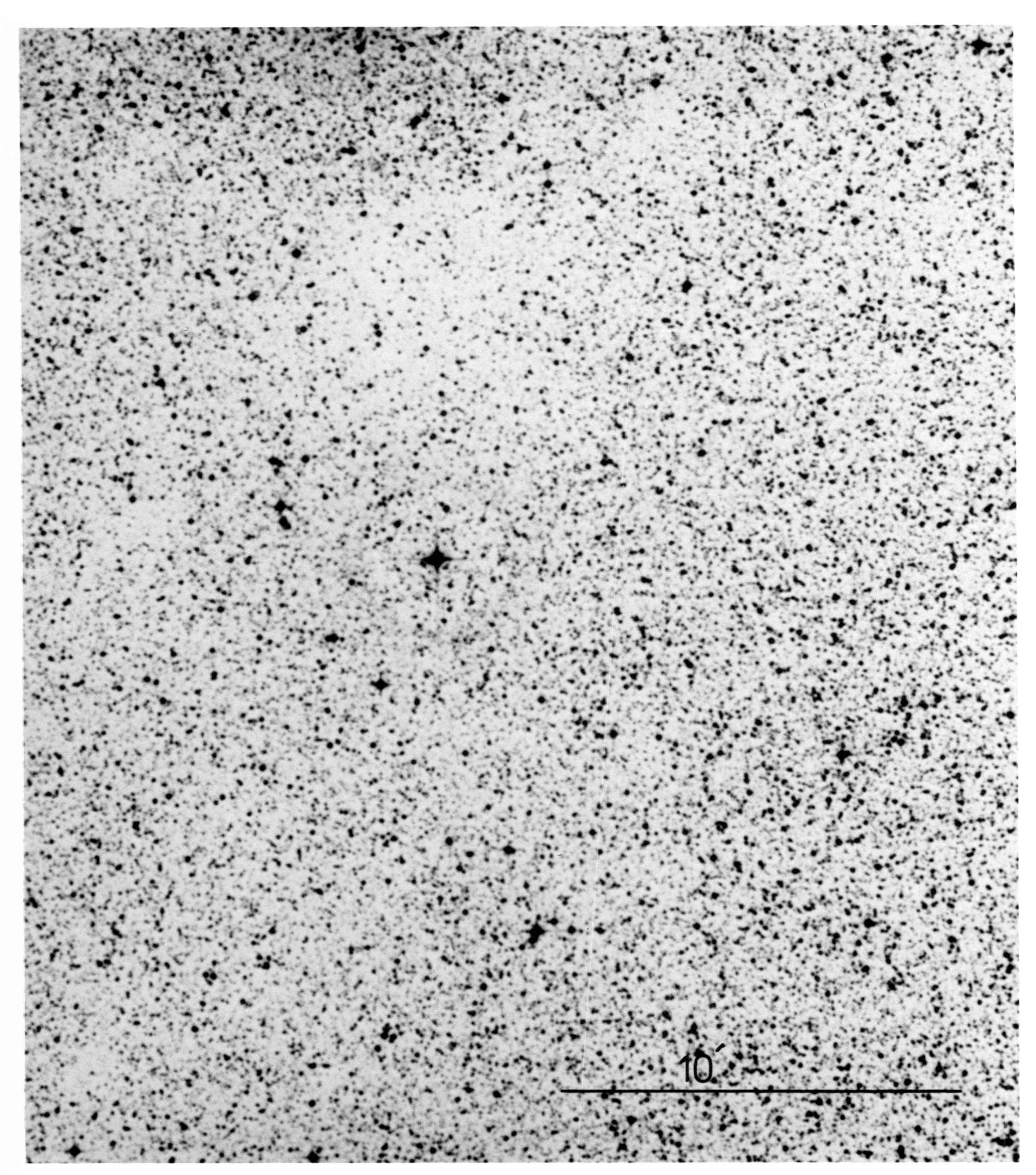

Plate 8.3

Repeat the observations using the identical regions and window for the other photographs. Compare the extinctions in the three wavebands. Try how well they approximate to the $1/\lambda$ law by plotting the extinction in magnitudes against the inverse of the wavelengths.

THE DENSITY OF GRAINS IN THE LINE OF SIGHT

Given the dimensions of the grains and their scattering properties it is possible to calculate the density of grains in a dark cloud. The effect of scattering may be described in the following way. If a grain is a sphere of radius a its cross-sectional area in the path of the light is πa^2. When scattering takes place the amount of light removed by the grain is $\pi a^2 Q$ where Q depends on the wavelength and the size of the grain. For interstellar grains, at short wavelengths Q is greater than one and the grain behaves as if it were larger than its actual geometrical size. For long wavelengths Q is less than one and the grain behaves as if it were smaller than its actual size.

For our estimate of the numbers of grains causing the extinction we assume that in the blue waveband the grains behave approximately like simple obstructions in the path of the light from the stars behind them (figure 8.2). The remainder of the light (apart from a small fraction which is absorbed by the grains) passes through in the spaces between the grains. The amount of light intercepted per unit area (as seen in projection on the sky) is proportional to the number of grains per unit area in the line of sight. If there are n grains per unit area projected on the sky, their combined area is $\pi a^2 n$ and the unobscured area is $(1 - \pi a^2 n)$ of a unit. The ratio of the light which gets through to the original light is $(1 - \pi a^2 n)$.

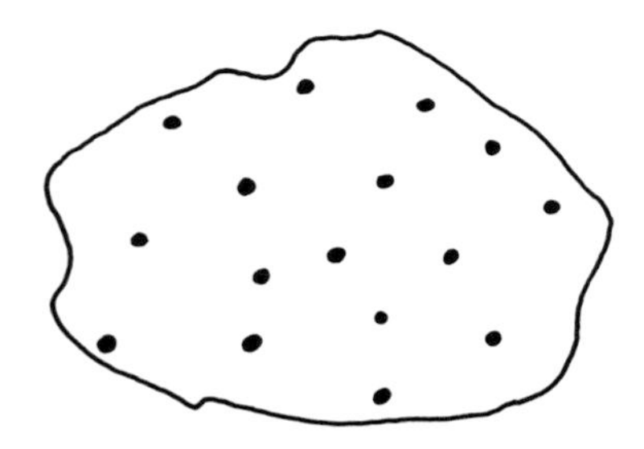

Figure 8.2 Cloud of grains projected on the plane of the sky. In this simple model the grains block out light from the stars behind them in proportion to their combined cross-sectional area.

Expressed in magnitudes, the extinction is 2.5 times the logarithm of this quantity (with the sign changed according to convention). In algebraic form

$$\text{extinction } \Delta m = -2.5 \log (1 - \pi a^2 n) \text{ mag.}$$

Exercise 2. Using the amount of extinction (in magnitudes) in blue light calculated in Exercise 1, estimate the number of grains per unit cross-sectional area in the cloud, given that the radius of a grain is 3×10^{-7} m. If the cloud is at a distance of 200 pc find its dimensions from the photograph and estimate the total number of grains in the cloud. The scale of the photographs is shown on plate 8.3.

Suggestions. Substitute the extinction Δm and the radius of a dust grain in the formula for Δm and solve for n, the number of grains per unit area of the obscuring cloud in the plane of the sky. Since the size of the grains is given in metres, the answer will be in numbers per square metre. You now need to find the cross-sectional area of the cloud in order to know how many grains there are altogether. If the distance to the cloud is r metres, and its surface area on the sky is A sr (steradians or square radians), the area of the cloud is Ar^2 m^2. You now simply multiply the number per square metre by this area to get the total number.

To get the area of the cloud in steradians you must first get its area on the photograph; one way is to draw a patch of the same size on millimetre graph paper and to count squares, another is to treat the cloud as approximately round and measure its radius. The scale of the photograph is given; convert the area first to square arcminutes and then to steradians (see Appendix 1). This is a simple calculation which does not need to be explained in detail; however, be careful with units.

SOLUTIONS AND DISCUSSION

1. A set of counts inside and outside the cloud is given in table 8.1. It should be emphasised that the actual counts vary from one observer to another; however, the ratios ought to be the same within observational errors (in this case ± 0.2 magnitude). Wavelengths are in microns (1 micron $= 10^3$ nm) for convenience. A rate of increase of 0.40 in log N per magnitude has been used, which means that Δm is 2.5 log (N/N_0). In fact, the exact value of the constant of

Facing page:

Plate 8.3 The same field as plate 8.1 in near-infrared (I) light. The plate scale is indicated (© Royal Observatory, Edinburgh).

proportionality does not matter in this instance: the purpose is to demonstrate that Δm is proportional to $1/\lambda$. The results are plotted in figure 8.3. Naturally, three points are rather few, but are enough to show how distinctly extinction increases with shorter wavelength.

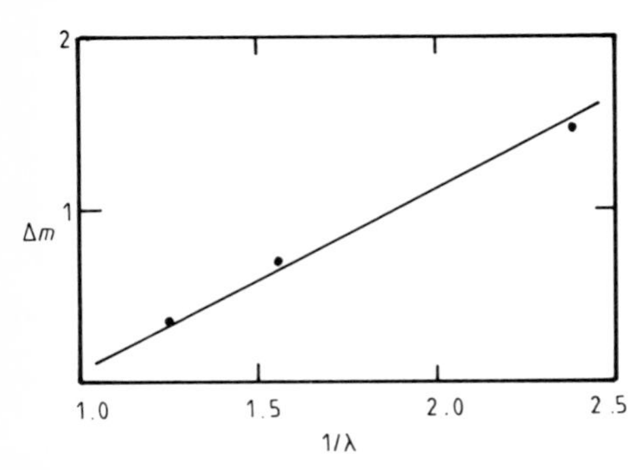

Figure 8.3 Plot of extinction in magnitudes against the inverse of wavelength in microns.

Table 8.1

λ (microns)	1/λ	N (inside)	N_0 (outside)	log (N/N_0)	Δm
0.80	1.25	58	75	−0.11	0.28
0.64	1.56	28	52	−0.27	0.67
0.42	2.38	25	100	−0.60	1.50

Detailed study of the problem makes use of spectroscopy in which several individual wavelengths are observed from infrared to the far-ultraviolet, recorded from satellites. The extinction curve over this entire range is shown for interest in figure 8.4, where it is seen that it is a straight line in the range which we have studied in the exercise, but diverges greatly from a straight line in the ultraviolet part of the spectrum.

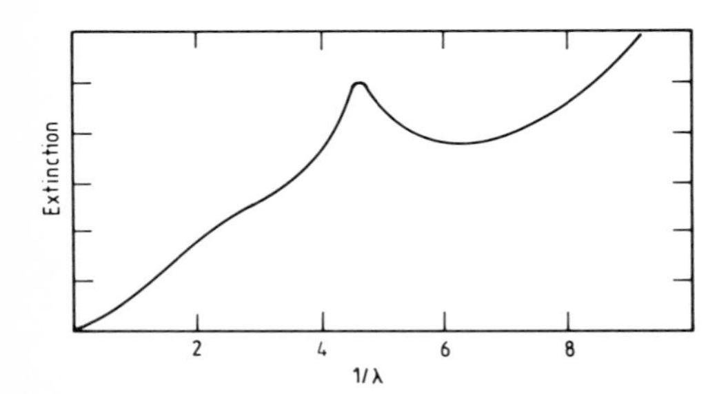

Figure 8.4 Standard extinction curve, from infrared to ultraviolet. It is seen that in the range of wavelength observed in the exercise the curve is a straight line.

2. Extinction is 1.5 magnitudes, therefore $(1-\pi a^2 n)=10^{-0.6}=0.25$.

$$\therefore\ \pi a^2 n = 0.75 \text{ or } n = 0.75/\pi a^2 = 2.6\times10^{12} \text{ grains per square metre.}$$

The diameter of the cloud is approximately 5 arcmin or $\frac{1}{12}$ of a degree.

$$\therefore \text{ the area} \cong \pi(1/24)^2 \text{ square degrees} = 1.6\times10^{-0.6} \text{ sr.}$$

The distance is 200 pc $= 6\times10^{18}$ m,

$\therefore$ the area $= 6\times10^{31}$ m^2 and the total number of grains $= 1.5\times10^{44}$.

Note that it does not matter whether the grains are all arranged in one plane like a sheet or are at different depths in the line of sight.

One could go even further and estimate the mass of the cloud of grains. The mass of a grain is its volume times the density of the grain material. A grain of silica or graphite of the size which we have assumed, has a mass of about 2.5×10^{-16} kg. Multiplied by the total number of grains this gives a mass of 4×10^{28} kg. This mass can best be visualised by comparing it with the mass of the Sun, 2×10^{30} kg. Dust is always accompanied by gas in the interstellar medium. Gas (mainly hydrogen atoms) is 10 times more abundant by mass than the dust; thus the total mass is about 10 times higher than the figure calculated for dust alone.

The results are very approximate, because of the limitations of the observations and the simplicity of the model. Experimenters who are used to laboratory work may think that the uncertainties are very great. Astronomers, on the contrary, often have to be satisfied with an 'order of magnitude' result, that is, a figure which is correct within a factor of about ten. In this present example, the total mass of the cloud may be described as of the order of $\frac{1}{10}$ of a solar mass.

Foreground star numbers in the field. We now confirm what we assumed earlier, i.e., that the stars seen in the dust patch are all likely to be in the background. The only information needed is the density of stars in nearby space. This is a well known quantity; it is (in round figures) 1 star per 10 pc^3.

The volume of space out to 200 pc is $\frac{4}{3}\pi(200 \text{ pc})^3$. Multiply by 0.1 to get the number of stars $= 3.3\times10^6$.

Divide by 4×10^4 (the number of square degrees on a sphere) to get the number per square degree. Divide by 100 (because the area of the cloud is about 1/100 square degree) to get the number in front of the cloud. The result is 0.8 stars. (In fact the number is smaller because the least luminous stars are not seen to a distance of 200 pc.) Hence it is justified to assume that all the stars seen at the cloud are behind it.

Correcting magnitudes and colours for extinction. From the established extinction law the relative extinctions at any desired wavebands are known. At the standard wavebands V and B the extinctions are in the approximate ratio of 3 : 4; that is, if the extinction in V is 3 mag, the extinction in B is 4 mag and the colour index $(B-V)$ is increased by one magnitude. The increase in colour, that is, the difference between observed and intrinsic colours, is called the *colour excess* or the *reddening* (written E_{B-V}). The way to allow for extinction in the magnitude of a star is to find first its colour excess by discovering its unreddened or intrinsic colour (for example from its spectrum) and subtracting it from its observed colour. The colour excess is then multiplied by three to give the extinction in the V waveband (written A_V)

or by four to give the extinction in the B waveband, A_B:

$$A_V \cong 3E_{B-V}$$

(more exact values of 3.1 and 3.2 have been determined by various observers for the coefficient).

In the case of star clusters, in which normally all stars are reddened by the same amount, the usual method of finding the extinction is to observe two colour indices, $(B-V)$ and $(U-B)$, for all the stars. (These indices are defined in Appendix 2.) There is a well known relation between the intrinsic values of these two colour indices for main sequence stars, and it is possible therefore to deduce by how much the observed colours need to be decreased in order to recover that relation. That decrease is the colour excess. The extinction in the star cluster NGC 2477, examined in §6, was obtained by this method. The colour excess E_{B-V} in that cluster is 0.53 mag, hence $A_V = 3.2 \times 0.53$ or 1.7 mag, the figure quoted in the Discussion of that chapter.

9 A Supernova and Supernova Remnants

The name *nova* (plural novae), an abbreviation of the Latin *stella nova* meaning 'new star', is the name given to a star which undergoes a sudden considerable increase in brightness which in pre-telescopic days was thought to signify the appearance of a new star where none had previously existed. The sudden outburst is caused by material ejected from the star into space.

Systematic searches for novae in external galaxies carried out in the 1930s at the Mount Wilson Observatory in California revealed that some novae were in an entirely different class from others, with luminosities equivalent to the luminosity of an entire galaxy. These amazing objects whose sources of energy were a complete mystery at the time were dubbed *supernovae*.

It is now known that a supernova marks a massive explosion caused by the death of a certain type of star during which most of the material of the star is blown out with immense energy. The explosive motion at 10^7 m s^{-1} or more is recognised from the Doppler shift in the spectrum of the ejected material. The expanding shell remains visible as a bright ring in interstellar space for thousands of years before fading eventually into the general interstellar medium. The visible vestige of a supernova is called a *supernova remnant*. The supernova leaves behind a residue in the form of a highly compact object of extreme density known as a neutron star. Some neutron stars become black holes. Such dense objects were predicted theoretically before they were actually observed. Their existence was demonstrated with the important discovery by Cambridge radio astronomers in 1968 of pulsars, objects which emit very rapid pulses of energy, interpreted as emerging from rapidly rotating neutron stars. Pulsars have sometimes been found where the visible remnants of a supernova have faded or are hidden by interstellar dust. The pulsars themselves eventually slow down and cease to radiate.

Supernovae divide into two classes, types I and II. Type I supernovae reach a maximum brightness with absolute magnitude -19 (equivalent to billions of Suns) in a matter of days, and then fade, first fairly quickly and then more slowly, over a period of some hundreds of days. Supernovae of type II are somewhat less luminous at maximum brightness and less regular in their rate of fading. The differences between the two types are interpreted as caused by differences in the progenitor or pre-explosion stars. Type II supernovae are the inevitable end product of massive young stars; most supernovae belong to this type. The progenitors of type I supernovae, on the other hand, are old stars of low mass (or higher mass stars which have suffered substantial mass loss) which have undergone an unusual evolution as members of a binary (double star) system. Normal stars of low mass, like the Sun, end as white dwarf stars which slowly fade away when all their nuclear energy has been used up. In the case of type I supernovae it is believed that the star in the last stages of its evolution receives extra material from its companion star which increases its mass above the limit of 1.4 solar masses, which is the maximum possible mass of a white dwarf star.

The star's later evolution then follows that of the more massive ones ending as a neutron star.

In our Galaxy, and also in other galaxies, as mentioned in §4, the description 'population I' is given to young stars which are found in the spiral arms, while the description 'population II' is given to older stars in the galactic halo. It is worth remarking that type I supernovae belong to population II and that type II belong to population I.

Only three supernovae have been recorded in our Galaxy in historical astronomical annals—one recorded by Chinese astronomers in the constellation of Taurus in 1054 AD; one in the constellation Cassiopeia observed by the great astronomer Tycho Brahe in 1592, and one in the constellation Ophiuchus, observed by another great astronomer, Kepler, in 1602. Tycho Brahe and Kepler noted the magnitudes of their supernovae, a valuable record which enables us to calculate the absolute luminosity in each case, once the distance to the associated pulsar has been estimated by the methods of radio astronomy.

Further knowledge comes from observations of supernovae in external galaxies. The rate at which supernovae occur is estimated at two per century per galaxy. This is not a large number from our observational point of view, firstly because most galaxies are very distant and also because supernovae occur without warning so that there is an element of luck in their discovery. The majority of supernovae which occur in our own Galaxy go unrecorded because of interstellar obscuration in the galactic plane which prevents us from seeing objects, even those as luminous as supernovae, in distant parts.

A record of the rise and fall of luminosity in a variable star is called its *light curve*. Figure 9.1 shows the shape of the light curve of a type I supernova.

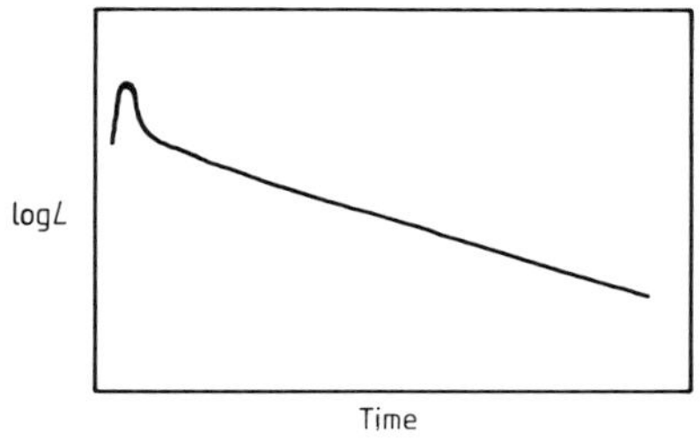

Figure 9.1 Typical shape of the light curve of a supernova.

SUPERNOVA 1987A

One of the most exciting events in the history of astronomy was the appearance of a supernova, visible to the naked eye in the Large Magellanic Cloud on February 23, 1987. The supernova was labelled 1987A, being the first observed in that year.

That the supernova should occur in the Large Magellanic Cloud was extremely fortunate for astronomers. The Large Magellanic Cloud is the nearest of all external galaxies, at a distance of only 50 kpc, a distance which is accurately known. It is situated at high galactic latitude and is therefore almost entirely free from interstellar obscuration. The supernova was observed almost from the moment of its occurrence, and for the first time in supernova records, the particulars of the original star prior to its eruption are known. For the first time also, the switching on of the pulsar in the remnant has been observed.

The progress of the supernova was recorded photographically by the UK Schmidt Telescope. Since the object was very bright, exposure times were short compared with those required for normal survey photographs. The series of photographs (plate 9.1) taken on the dates listed in table 9.1 shows the supernova and the nearby huge bright nebulosity, 30 Doradus (Dorado is the constellation of the Swordfish in which the Large Magellanic Cloud is situated). You can find your bearings by identifying 30 Doradus on the photograph of the Large Magellanic Cloud (plate 11.3). The progenitor star of the supernova, a blue star of magnitude 12.2, is present on plate 9.2, which was taken before the supernova outburst, as you may see by looking carefully at it.

Pages 62–3:

Plate 9.1 A series of 12 photographs of supernova 1987A. The dates on which they were taken are given in table 9.1 (© Royal Observatory, Edinburgh).

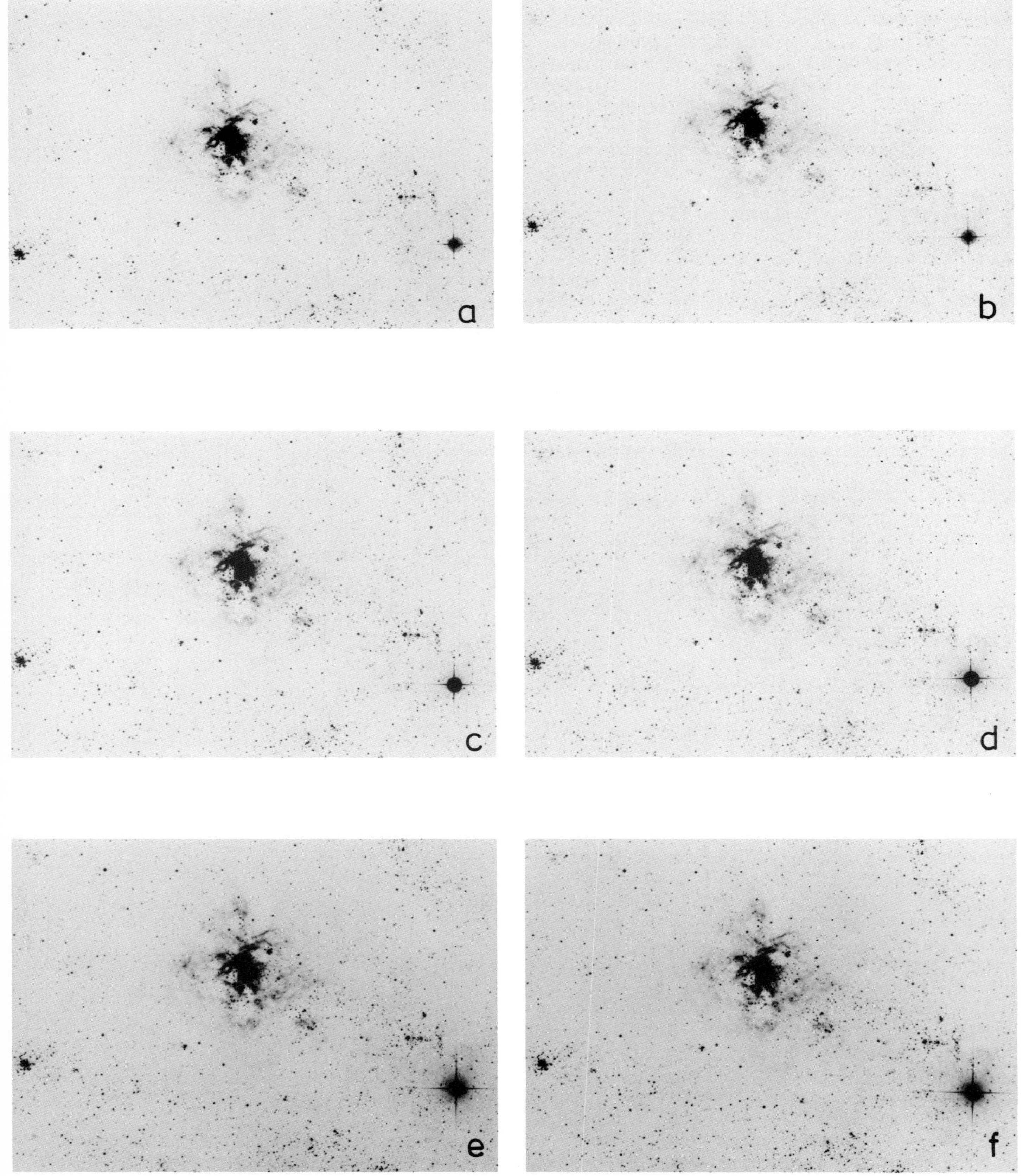

Plate 9.1

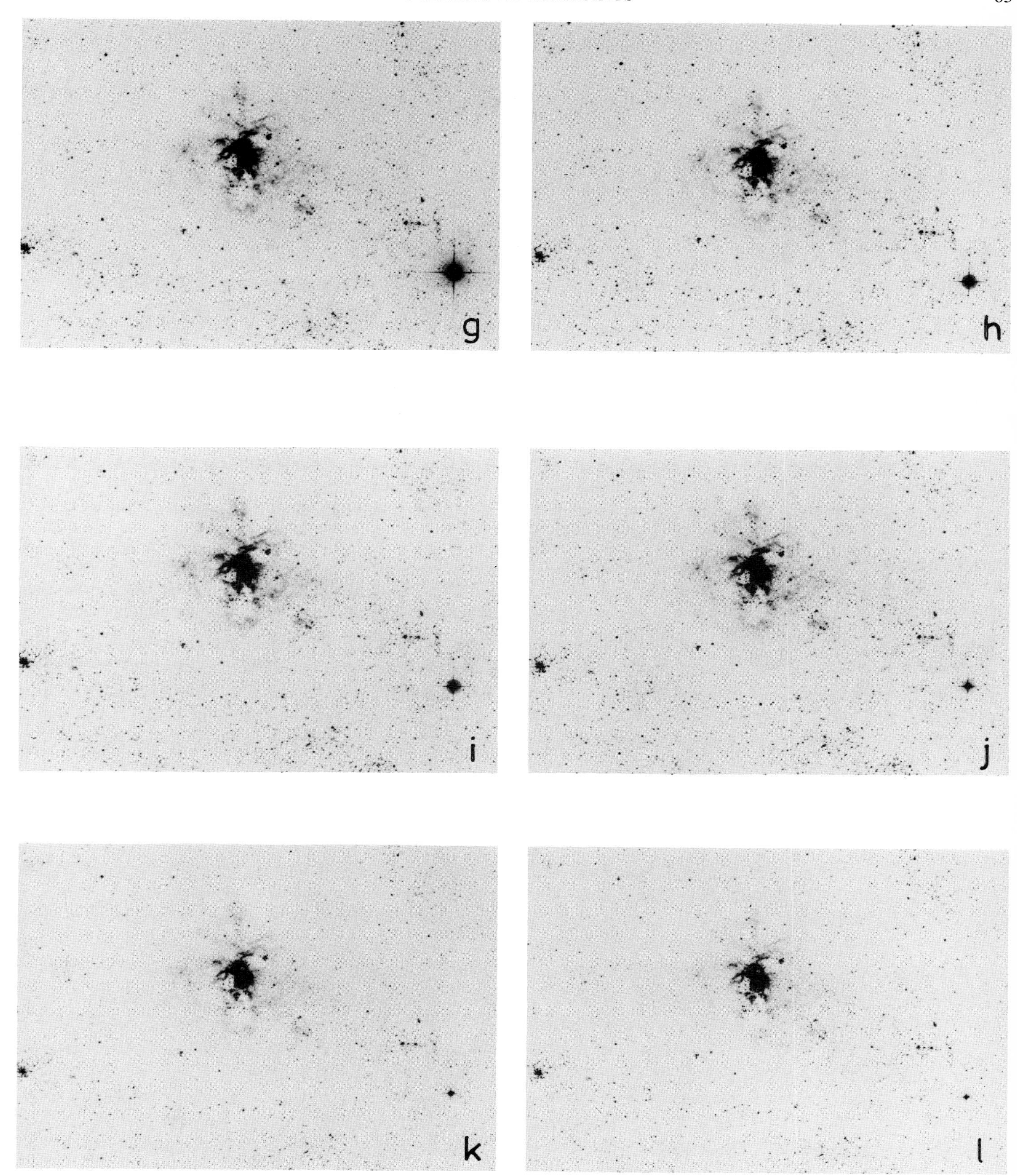
g
h
i
j
k
l

Table 9.1 Dates of the photographs of SN 1987A.

a	1987	27 February
b		10 March
c		29 March
d		8 April
e		24 April
f		5 May
g		25 May
h		4 July
i		15 August
j		17 October
k		9 December
l	1988	11 February

We have seen (§4) that photographic star images require calibration if the magnitudes of stars are to be known. No calibration exists for the set of photographs of the supernova because the surrounding star field contains no star bright enough to match the supernova during the greater part of its appearance. However, even without calibration, it is possible to trace the rise and fall in brightness of the supernova from the varying size of its image on the photographs. We begin by assuming that the photographs are otherwise identical—a reasonable assumption which is to be tested later—and by recalling that calibration curves, as shown in §4, are fairly linear over a considerable range of magnitude. The measured diameter of the supernova image is thus assumed to be closely proportional to its magnitude. Unfortunately the bright images of the supernova are surrounded by a

Plate 9.2 The field of the supernova photographed before the supernova's occurrence (© Royal Observatory, Edinburgh).

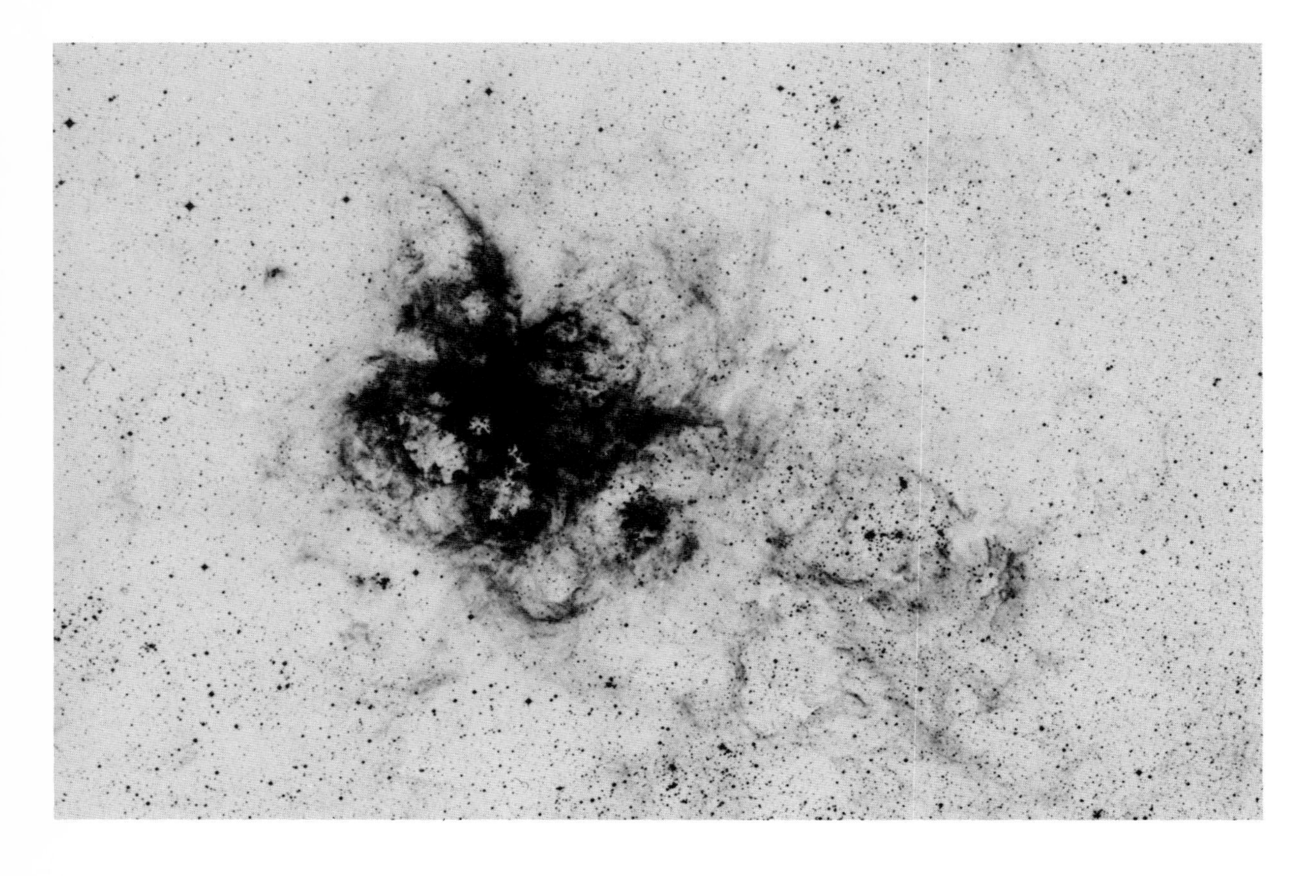

halo. This is an instrumental effect which tends to mask the true image of the star which, however, is distinguishable on careful examination.

To test the assumption that the changes in the supernova shown in the photographs are solely in the supernova itself and not due to instrumental or observational effects, one or more control stars on each photograph should be observed to monitor such possible effects.

Exercise 1. Draw a light curve of Supernova 1987A using image diameter as an indicator of magnitude.

Suggestions. Measure the diameter of the image of the supernova on each photograph to the nearest 0.1 mm. Convert the date on each photograph to days, counting from Day 0, the date of the first photograph (or, if you prefer, the date of the discovery, February 23). Plot the diameter against time in days. The plot will show the contrast between the rate of rise and the rate of decline of the supernova. From your experience in measuring the diameters you will have an idea of how much uncertainty there is in the plotted points. Draw a mean smooth curve through the points, allowing for this scatter. On account of the intervals between photographs, the exact instant of maximum brightness may not be recorded.

Measure also the diameters of the monitor star or stars on each photograph, and plot these measurements against date on the same scale as the light curve. Again there will be scatter in the points, due to your errors of measurement. Only a variation larger than the expected scatter and showing up on all the monitored stars would be significant. Such a significant variation, if found, would have to be taken into account in assessing the brightness of the supernova on that particular date.

Exercise 2 (numerical calculation). The magnitude of the supernova at maximum brightness was $V = 3.0$. Calculate the luminosity of the supernova in solar units, given the distance of the Large Magellanic Cloud (50 kpc) and the absolute V magnitude of the Sun, +4.8 mag, and also in watts, given that the Sun's luminosity is 4×10^{26} W. Assume in this calculation that the supernova has the same colour as the Sun, i.e., radiates as the Sun but on a larger scale. Refer if necessary to the definitions of magnitude and luminosity in Appendix 2.

THE CRAB NEBULA

The Crab nebula (plate 9.3), number 1 in Messier's list of unusual objects (M1, NGC 1952), at coordinates (05h 34m, +22.0°) in the constellation Taurus, is the remnant of the supernova which was observed by Chinese astronomers in the year 1054 AD. The Chinese astronomers recorded the exact date, July 4, of the first appearance of the 'guest star' which was as bright as the planet Venus and visible in daylight. It remained very bright for 23 days, and took two years to fade completely. It is one of the puzzles in the history of astronomy that this remarkable spectacle in a well known constellation went unrecorded by western observers.

The Crab nebula is so called from its appearance, with its claw-like boundaries reminiscent of a crab. The realisation that the Crab nebula is an expanding mass of gas came about from observations early in this century of Doppler shifts in its spectrum.

The strong radio source, Taurus A, first recorded in 1949, was found to coincide with the Crab nebula, and in 1969 one of the newly discovered pulsars was also identified with it. The Crab pulsar has a period of 1/30 of a second and the residue of the actual star, now a neutron star, is still observable in optical light inside the nebula. In the supernova explosion, material from the star was thrown into the surrounding space. The present tangled appearance represents this expanding material, emitting radiation which is identified from its spectrum with *sychrotron radiation*, that is, radiation caused by fast moving electrons following spiral paths around lines of magnetic field. Radiation emanating from material falling on the surface of the neutron star is directed in a narrow beam which causes it to be observed in pulses as the beam periodically points towards us from the fast rotating neutron star.

The outermost part of the shell of the Crab nebula emits strongly the red spectrum line (Hα at 656 nm) characteristic of hydrogen. In this case the emission at the periphery of the nebula originates in interstellar space. The

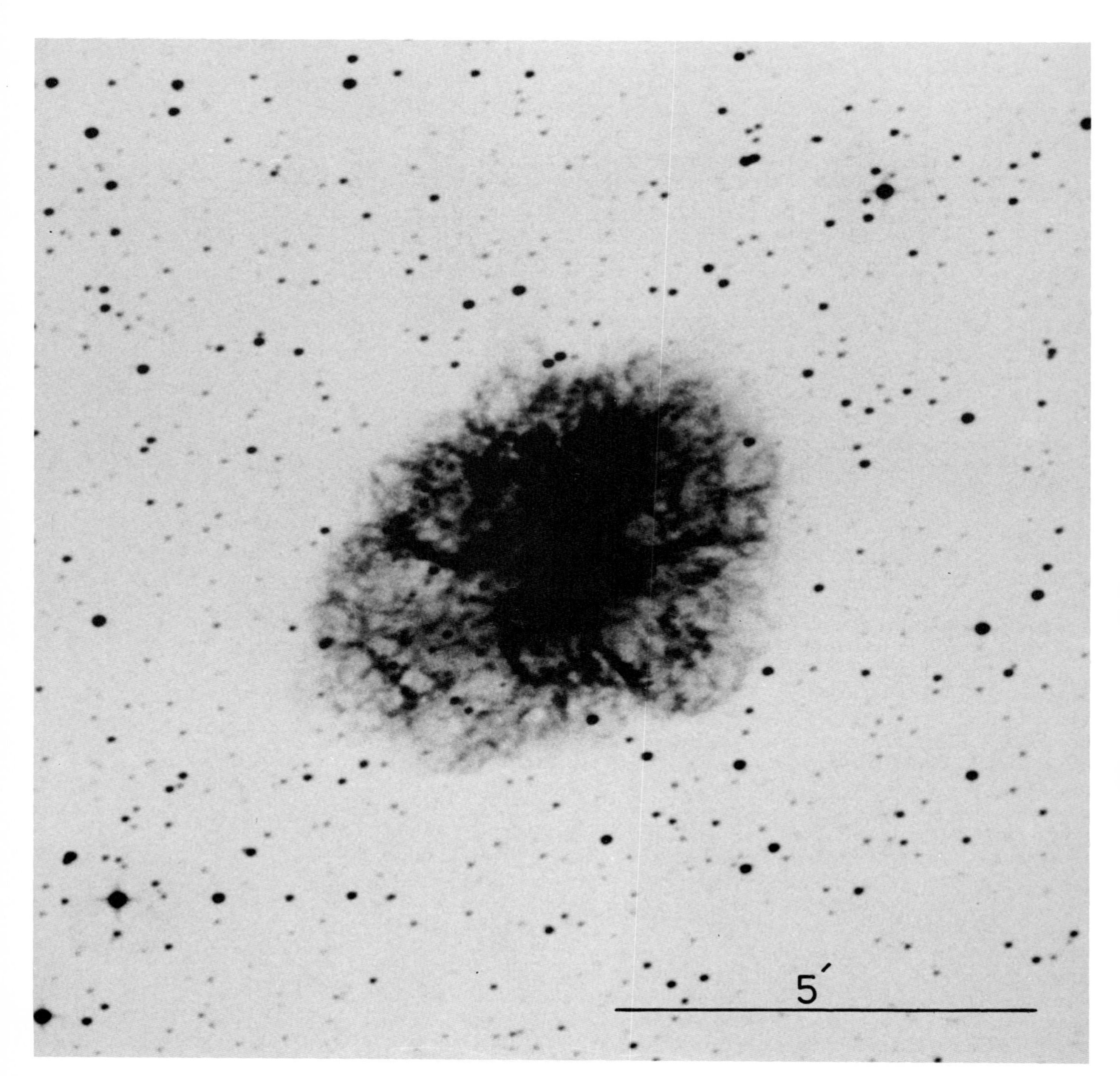

Plate 9.3 The Crab nebula, a supernova remnant. The pulsar (not marked) is inside the nebula (© Royal Observatory, Edinburgh).

expanding material from the explosion has pushed the surrounding interstellar material in front of it; this interstellar material which now makes up the outer shell of the Crab nebula is mainly composed of hydrogen in ionised form (H II). What is more, the velocity in the line of sight of the H II shell measured spectroscopically by the Doppler effect shows a velocity of 1.5×10^6 m s^{-1} towards us (this is the material in the front part of the nebula in our line of sight) and the same velocity of recession, identified as originating in the opposite part of the shell. This velocity is therefore a velocity of expansion of the outer skin of the Crab nebula.

The expansion is also related to the present angular size of the Crab nebula and its age. If R is the radius of the Crab nebula in metres and t is its age in seconds, then the average velocity of expansion since the explosion began is R/t m s^{-1}. If the nebula expands uniformly in all directions and if the rate of expansion has remained constant since the supernova exploded, then this velocity of expansion must be the same as that observed from the line-of-sight motion. From this the radius R in metres of the nebula may be calculated. The present radius of the nebula in arcseconds is observed on photographs such as the one reproduced here. From these data it is possible to deduce its distance.

The photograph of the Crab nebula was taken through an Hα filter, that is, a narrow-band colour filter which transmits the Hα line of hydrogen at 656 nm and thus reveals clearly the outer boundaries of the nebula.

Exercise 3. From the observed dimensions of the Crab nebula on the photograph (taken in 1987) find its average annual angular rate of expansion from its beginning in 1054 AD. *Given the radial velocity of the outermost filaments (1.5×10^6 m s^{-1}) deduce the distance of the Crab nebula.*

Suggestions. The nebula is not spherical; from its shape in the photograph it is evident that its expansion has not been uniform in all directions. Its depth in the line of sight is therefore not known, but it is reasonable to suggest that it is similar to the dimensions on the plane of the sky. Assume that it is a mean between the long and the short axes of the nebula as seen on the plane of the sky. Measure the longest and the shortest diameters on the photograph to the nearest tenth of a millimetre, taking care to measure to the very edges; take the mean, divide by two and convert to seconds of arc using the scale marked on the photograph. Divide by the interval in years to give motion in the plane of the sky in arcseconds per year. The formula for converting proper motion to transverse velocity, derived in §5, may now be used. Since the transverse velocity is equal to the radial velocity, the formula gives the distance to the nebula in parsecs.

THE VELA SUPERNOVA REMNANT

The photograph (plate 9.4), of a full 6.5° × 6.5° field taken with the UK Schmidt Telescope centred on coordinates (08h 38m, −44.8°) in the galactic plane in the southern constellation of Vela (the Sails) shows another example of a supernova remnant. The visible outlines are the projection of a three-dimensional shell of scooped-up interstellar material. To accentuate the visible structure, mainly composed of ionised hydrogen, the photograph, like the previous one of the Crab nebula, has been taken through a filter which transmits a narrow wavelength band centred on the strong Hα emission line of hydrogen.

Though the Vela remnant belongs to an earlier supernova explosion than the Crab, it is still active and relatively young. It is associated with a pulsar at coordinates (08h 33m, −45.0°) which is also a source of x-rays. The star at the same position is too faint to be detected in the normal way and is not visible on our photograph; it remained unobserved until 1977 when astronomers succeeded in recording optical radiation which indicated a star of magnitude 24. This was achieved by pointing the telescope (the Anglo Australian 4 m telescope) at the precise position of the pulsar, and by synchronising the observations with the radio pulsar flashes and integrating the light signal over many hours. The Vela and the Crab remnants—and now the Large Magellanic Cloud supernova—are the only pulsars for which optical counterparts have been observed.

The distance to a pulsar in the galactic plane may be estimated by the radio method of 'dispersion measurement'. Radio waves travelling through a plasma (a medium containing electrically charged particles) suffer a delay which is more marked at low than at high frequencies. The effect is

Plate 9.4 The Vela supernova remnant. The position of the associated pulsar is marked (© Royal Observatory, Edinburgh).

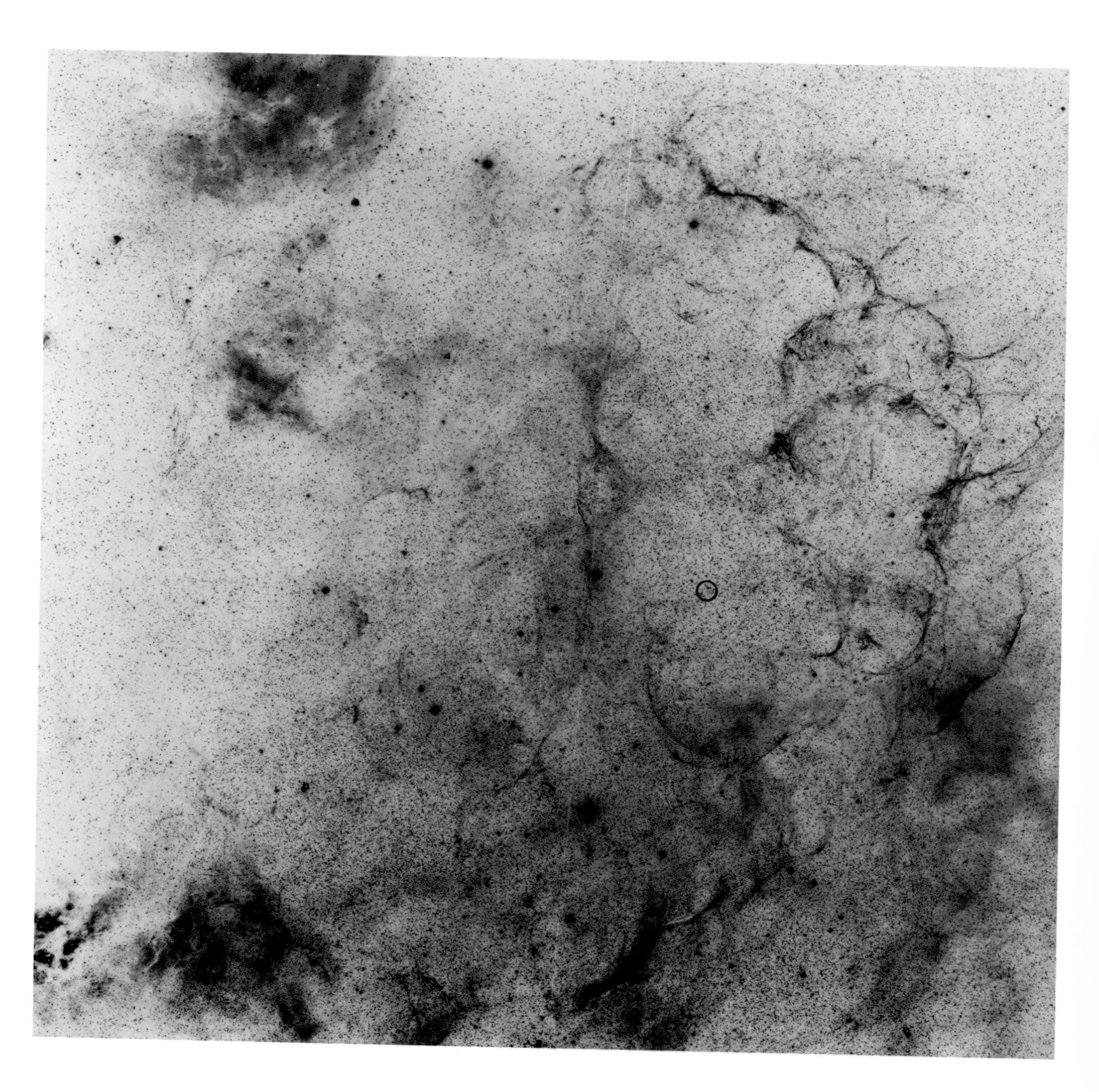

proportional to the number of electrons per unit area in the column of intervening space between the pulsar and the observer. If the density of electrons per unit volume is assumed constant, the number of electrons in the column is proportional to the depth of the column, that is, to the distance to the pulsar. Radio astronomers use this method to estimate the distances to pulsars. In the case of the Vela pulsar, the distance from dispersion measurement is 500 pc.

Exercise 4. Find the radius in parsecs of the Vela supernova remnant, given that its distance is 500 pc. Estimate also the age of the remnant, assuming the same physical conditions as in the Crab nebula.

Suggestions. Try to find the centre of the shell independently, without being influenced by the centering of the photograph or by the position of the pulsar which is marked on the photograph. Draw a number of concentric circles with a compass on a sheet of transparent paper or overlay. Place the paper on the page and move it around until you find the best position which shows the majority of filaments lying parallel to the circles. Mark the centre lightly on the photograph. It is a good idea to repeat this exercise a few times, turning the page upside down or sideways between attempts. Naturally there is an element of uncertainty because the shell is not a perfect sphere.

Having decided on the centre, measure the radius of the outermost part of the shell with a millimetre ruler. Convert to angular measure using the known dimensions (6.5° square) of the photograph and then to linear measure in parsecs assuming a distance of 500 pc.

The age and the present dimensions of the Crab nebula are known from the previous exercise. It is an easy matter to calculate the age of the Vela remnant on the assumption that both remnants expand at the same rate. All that is needed is to calculate how many times bigger the Vela remnant is than the Crab remnant, and to multiply the age of the Crab remnant by the same factor.

SOLUTIONS AND DISCUSSION

1. Figure 9.2 is a curve showing the diameter of the supernova image against date in days from February 23. The lower points, showing the variation of the diameter of a monitor star, demonstrate that there are no serious instrumental effects and that the curve represents genuine variations in the magnitude of the star.

Figure 9.3 is the observed bolometric light curve (i.e. the luminosity which includes light from all wavelengths); the resemblance of the photographic light curve to this real one is very satisfactory.

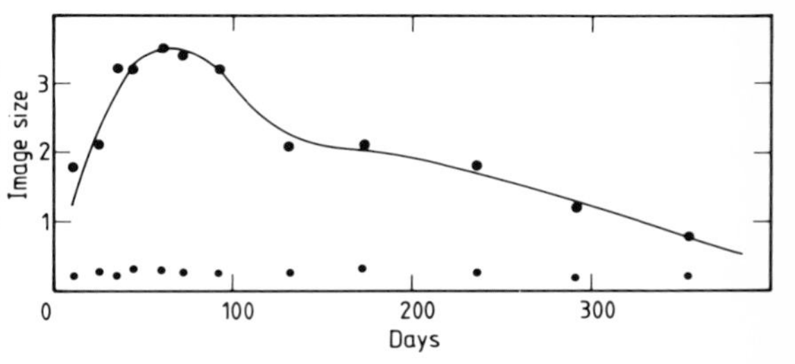

Figure 9.2 Light curve of supernova 1987A in the form of a plot of photographic image size against time in days.

2. The distance modulus of the supernova (5 log $(r/10)$ where $r = 50\,000$) is 18.5.

$\therefore$ absolute magnitude $= 3.0 - 18.5 = -15.5$
absolute magnitude of Sun $= 4.8$
difference $= -20.3$
$\therefore \log (L/L_0) = 0.4 \times 20.3 = 8.12$
$\therefore L/L_0 = 1.3 \times 10^8$
$\therefore L = 5.3 \times 10^{34}$ W.

The calculation assumes that the supernova has the same colour as the Sun, which is in fact not far from the case. The colour of the supernova was $(B-V) = 1.0$; that of the Sun is 0.8, a difference of only 0.2 mag.

An interesting characteristic of the supernova is its low luminosity. Its progenitor star was a blue supergiant star (absolute magnitude -6.4), a member of population I which would be expected to give rise to a type II supernova (absolute magnitude -17). The absolute magnitude, however, was only -15.5.

3. The longest and shortest diameters of the nebula are 380 and 296 arcsec.

Mean radius = 169 arcsec,
interval of time = 933 years,
$\therefore$ annual motion = 0.18 arcsec,
corresponding velocity $= 4.74 \times 0.18 \times 10^3 r$ m s^{-1} where r is the distance in parsecs; this is equal to the radial velocity of 1.5×10^6 m s^{-1}.
$\therefore r = 1500/(4.74 \times 0.18) \cong 1750$ pc.

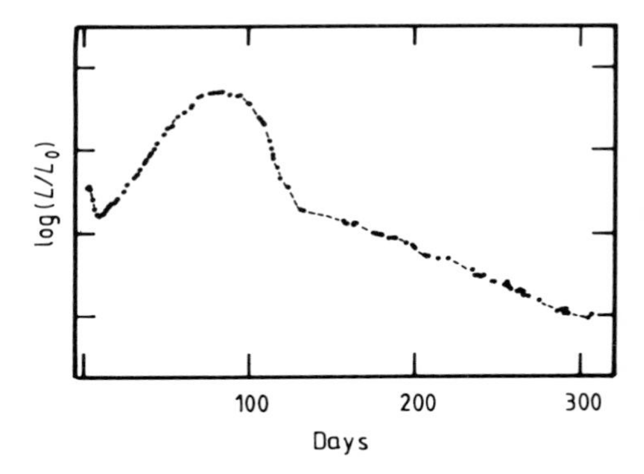

Figure 9.3 Actual light curve of supernova 1987A.

The uncertainty in this conclusion arises from two sources: the assumptions that the rate of expansion has remained constant from the beginning, and that the depth of the nebula in the line of sight equals the average of its other two dimensions. The first of these assumptions has been tested from observations of an actual increase in the dimensions of the Crab nebula as measured on photographs taken before and after an interval of 40 years. The radial increase observed is 0.21 arcsec per year. This result (which of course represents the expansion in recent times) and is slightly higher than our figure, has been interpreted as showing that there has been no slowing down of the expansion since it began. Uniform expansion of 0.21 arcsec per year gives a distance of 1500 pc.

4. The radius of the shell is about $2.3° = 0.040$ rad,

$\therefore$ linear radius $= 0.04 \times 500 = 20$ pc.
Linear radius of Crab shell $= 1.4$ pc,
$\therefore$ age of Vela remnant $=$ age of Crab remnant $\times (20/1.4) \cong 13\,000$ years.

The answer depends on the assumption that conditions in the two remnants were identical to begin with, and have remained identical. The age therefore is only an estimate. However, it agrees well with the age calculated by radio astronomers from the rate of slowing down of the pulsar, namely, 11 000 years.

The position of the pulsar does not coincide exactly with the centre of the shell. The reason for this is that the centre of the shell marks the position of the supernova when it first exploded; the pulsar marks the present position of what remains of the central star which may have recoiled according to the law of conservation of momentum as a result of asymmetry in the explosion. The observed separation of the pulsar from the centre of the remnant is the component of the pulsar's movement in the plane of the sky, and is thus a minimum value of its actual displacement.

10 Types of Galaxies

Galaxies are systems of stars external to our own. Before their true nature was known they were given the general label of *nebulae*, or clouds. Later, when it was realised that they are outside our Galaxy, they were called extragalactic nebulae, a confusing designation eventually replaced by the name of 'galaxies'.

The classification of galaxies into various types according to their appearance has many important purposes. One of these is their use as distance indicators. Galaxies are composed of vast numbers of stars and can therefore be seen to great distances. If a particular type of galaxy has its own unique characteristics, such as specific dimensions or luminosity, then a galaxy of the same type in a more distant part of the universe may reasonably be expected to have the same dimensions and luminosity. Apparent dimensions decrease inversely as the distance; apparent brightnesses decrease according to the inverse square law, that is, inversely as the square of the distance. For example, an object placed at twice the distance would appear twice as small but four times less bright. By observing the apparent diameter or the apparent brightness of a distant galaxy and comparing these with the same quantities in nearby galaxies of the same type, it is possible to estimate its distance. As galaxies vary greatly in dimensions and in luminosity it is important therefore to be able to recognise particular types of galaxy purely from their appearance.

CLASSIFICATION OF GALAXIES ACCORDING TO THE HUBBLE SCHEME

Galaxies are observed to be of broadly two kinds, spirals and ellipticals. The spiral shape was first recorded by Lord Rosse in 1846 with his 6 foot (1.8 m) reflector, then the world's largest. In 1924 Hubble subdivided spiral and elliptical galaxies in a classification scheme called the 'tuning fork diagram' (figure 10.1). In this scheme, elliptical galaxies are placed in a sequence of increasing flatness, from globular to lenticular (lens shaped). The types are labelled E0 (spherical) to E7. The spiral galaxies are subdivided along the prongs of the fork into those with a round central core (normal spirals labelled S) and those with an elongated central structure or 'bar' (barred spirals labelled SB). Within each division the galaxies are further placed in a sequence of increasing looseness of the spiral arms, from 'a' the most tightly wound to 'c' the loosest; the extent of their nuclei also decreases along the sequence. Hubble also added type S0 at the junction of the ellipticals and spirals in the diagram, though no examples of this type had been observed. This type, flat with minimal or no spiral arms, was later found to exist. Later, also, two other types were added: dwarf elliptical (dE) and dwarf spheroidal galaxies which are in shape and appearance like the regular ellipticals but are very small and of very low mass; and, at the other end of the sequence, a class called *irregular*. As their name implies, these are of irregular shape but have elements in common with spiral galaxies. Another category, discovered since Hubble's classification, are galaxies with dense bright cores, for example quasars, called N types.

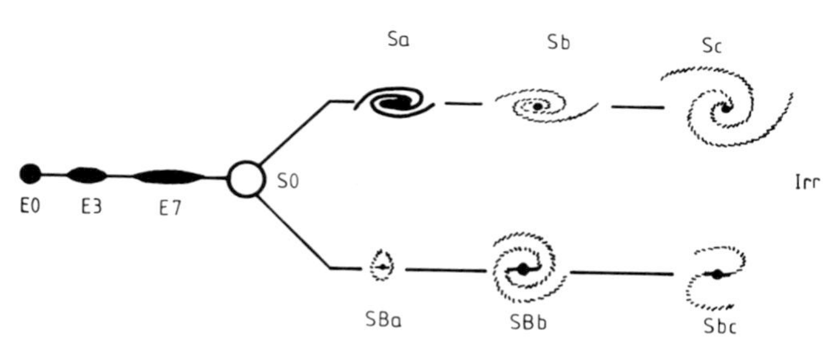

Figure 10.1 The Hubble tuning fork diagram for galaxy classification.

The Hubble sequence is one of stellar population type. Elliptical galaxies are composed of evolved stars, analagous to globular star clusters. Proceeding from left to right along the sequence, spiral galaxies show an increase in the luminosity of the arms, in blueness, in numbers of young stars and H II regions and in the proportion of gas. The sequence is therefore one of increasing youthfulness in the stellar population. From studies of our own Galaxy it is known that stars are formed in regions of high gas density in the spiral arms, that H II regions (containing ionised hydrogen and recognised spectroscopically) denote recent star formation, and that young luminous stars are blue. Galaxies are also increasingly more flat in shape as one goes from left to right in the sequence. Our Milky Way Galaxy is of type Sb. Spirals are in general more luminous than ellipticals, but there are some elliptical galaxies, the giant ellipticals, which are more luminous still. Large galaxies are surrounded by halos of globular clusters and are often attended by companion dwarf galaxies. Irregular galaxies account for only a few per cent of all galaxies.

The Hubble sequence has been refined and extended as observations have become more detailed, but the basic structure of the classification remains the same.

Galaxies tend to occur in groups or clusters. The first to note this was William Herschel who remarked that there were certain parts of the sky where the 'nebulae' came thick and fast. One of these was the constellation of Virgo in the northern sky. A look at a star atlas shows about 100 galaxies in a field of 12° diameter centred on coordinates (12h 27m, 13.5°). There are in fact many thousands of galaxies in the Virgo cluster which is the nearest large group of galaxies, at a distance of 20 Mpc (megaparsec).

Plates 10.1 and 10.2 show examples of individual spiral galaxies. Plates 10.4 to 10.7 show an extended area of the Virgo cluster in four parts and plate 10.3 shows the central galaxies of this cluster.

Exercise 1. Classify the spiral galaxies on plates 10.1 and 10.2 according to the Hubble scheme. Check your classifications with those given in table 10.1 before going on to classify the galaxies in the Virgo cluster on plates 10.3 to 10.7. The chart (figure 10.2) identifies the galaxies in the cluster by running numbers. There is a small overlap between the photographs to enable you to find your way on the chart. Other galaxies for classification are found on plates 4.1, 11.1, 11.3 and 11.4.

The Virgo cluster contains only normal galaxies; there are no N type or irregular galaxies. Galaxies are spread throughout space as well as in clusters, so the fields may contain galaxies which are more distant than the cluster. Near and distant spirals can be distinguished from their relative sizes but there is no way of distinguishing a nearby dwarf elliptical galaxy from a distant giant one from appearance alone. Some small elliptical galaxies, however, are recognised from their location as companions of larger galaxies.

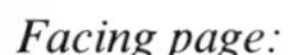

Facing page:

Plate 10.1 Examples of spiral galaxies (© Royal Observatory, Edinburgh).

Page 74:

Plate 10.2 More examples of spiral galaxies (© Royal Observatory, Edinburgh).

Page 75:

Plate 10.3 The central region of the Virgo cluster of galaxies (© Royal Observatory, Edinburgh).

Page 76:

Plate 10.4 The Virgo cluster of galaxies, north east quadrant (© Royal Observatory, Edinburgh).

Page 77:

Plate 10.5 The Virgo cluster of galaxies, north west quadrant (© Royal Observatory, Edinburgh).

Page 78:

Plate 10.6 The Virgo cluster of galaxies, south east quadrant (© Royal Observatory, Edinburgh).

Page 79:

Plate 10.7 The Virgo cluster of galaxies, south west quadrant (© Royal Observatory, Edinburgh).

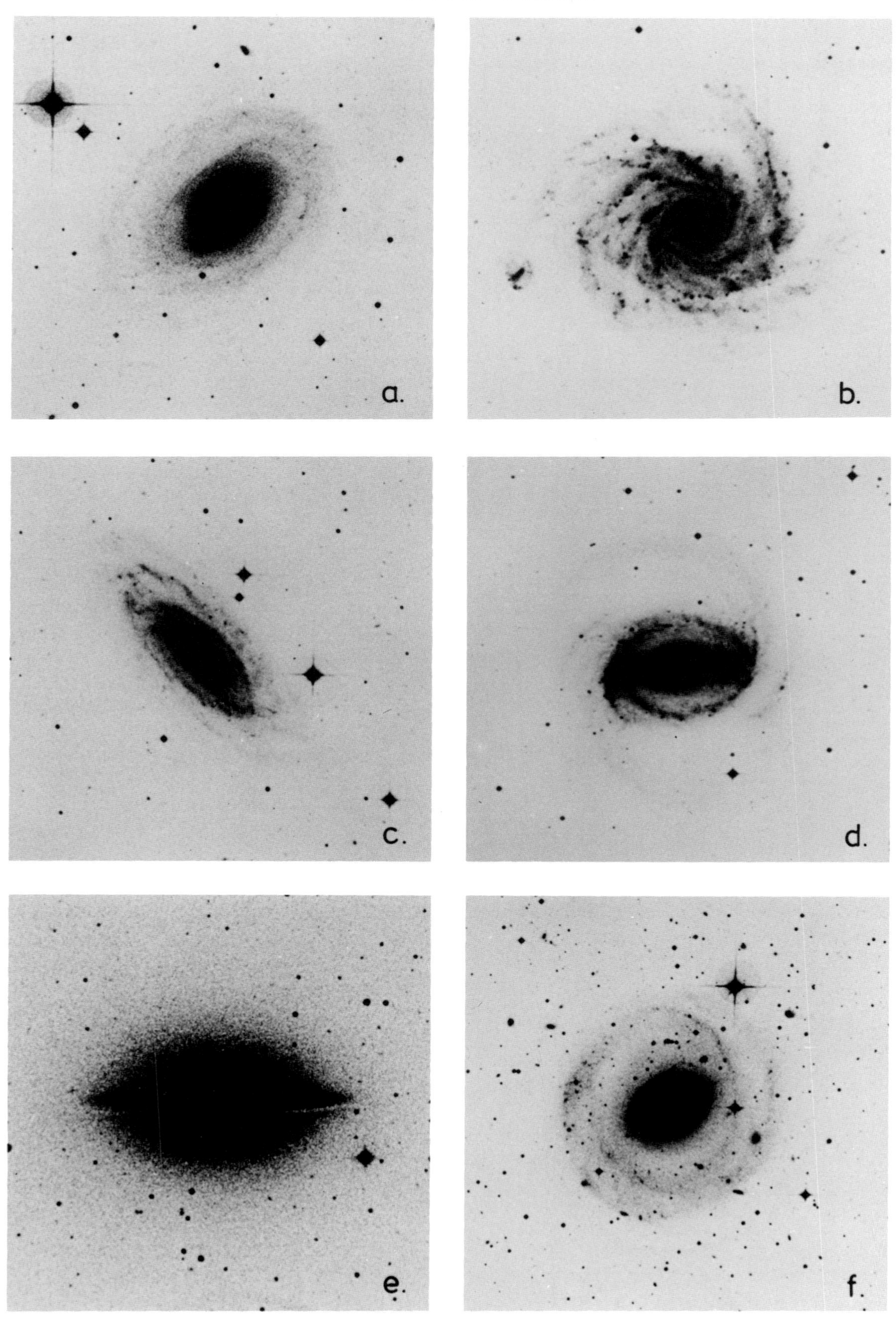

Plate 10.1

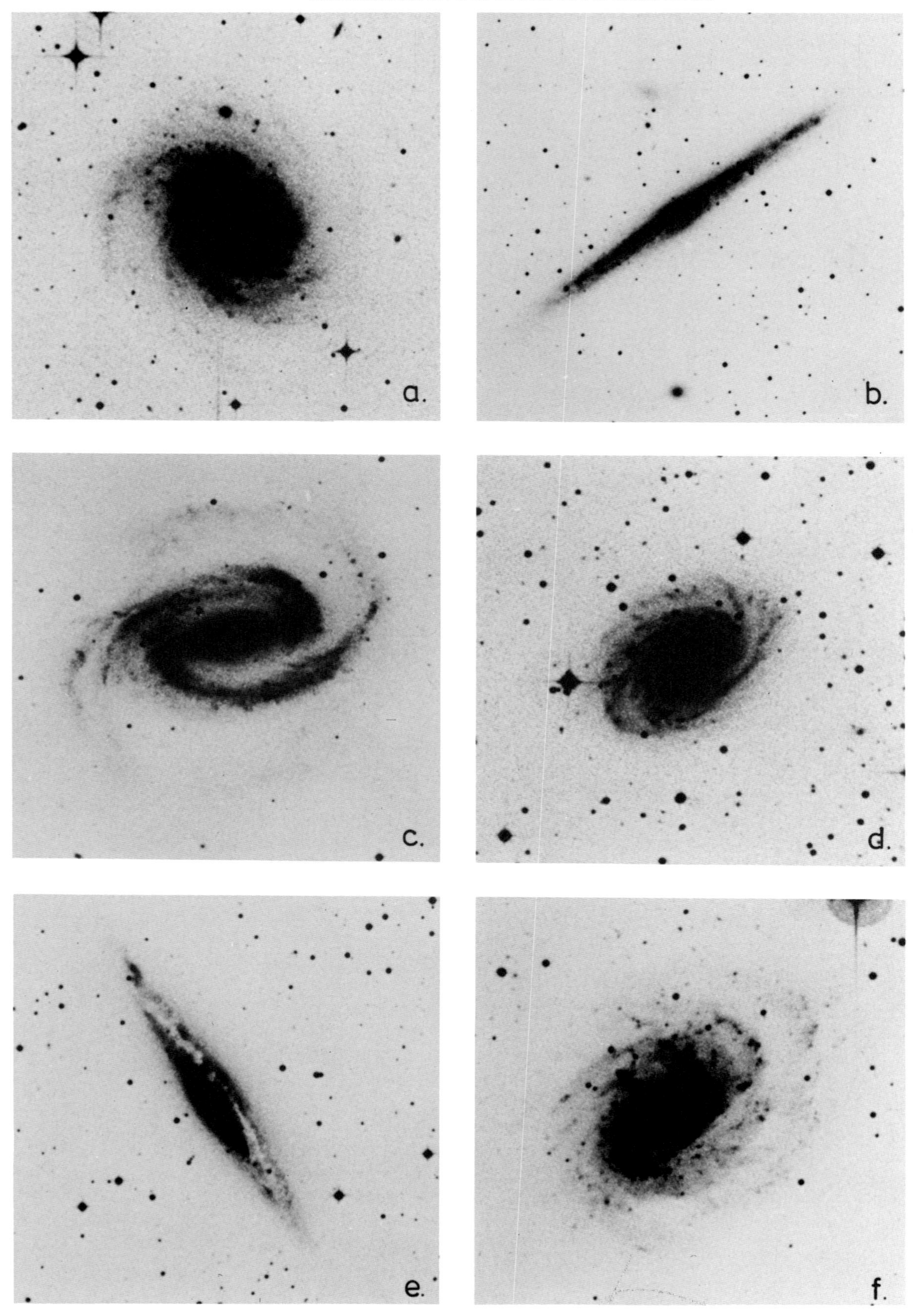

Plate 10.2

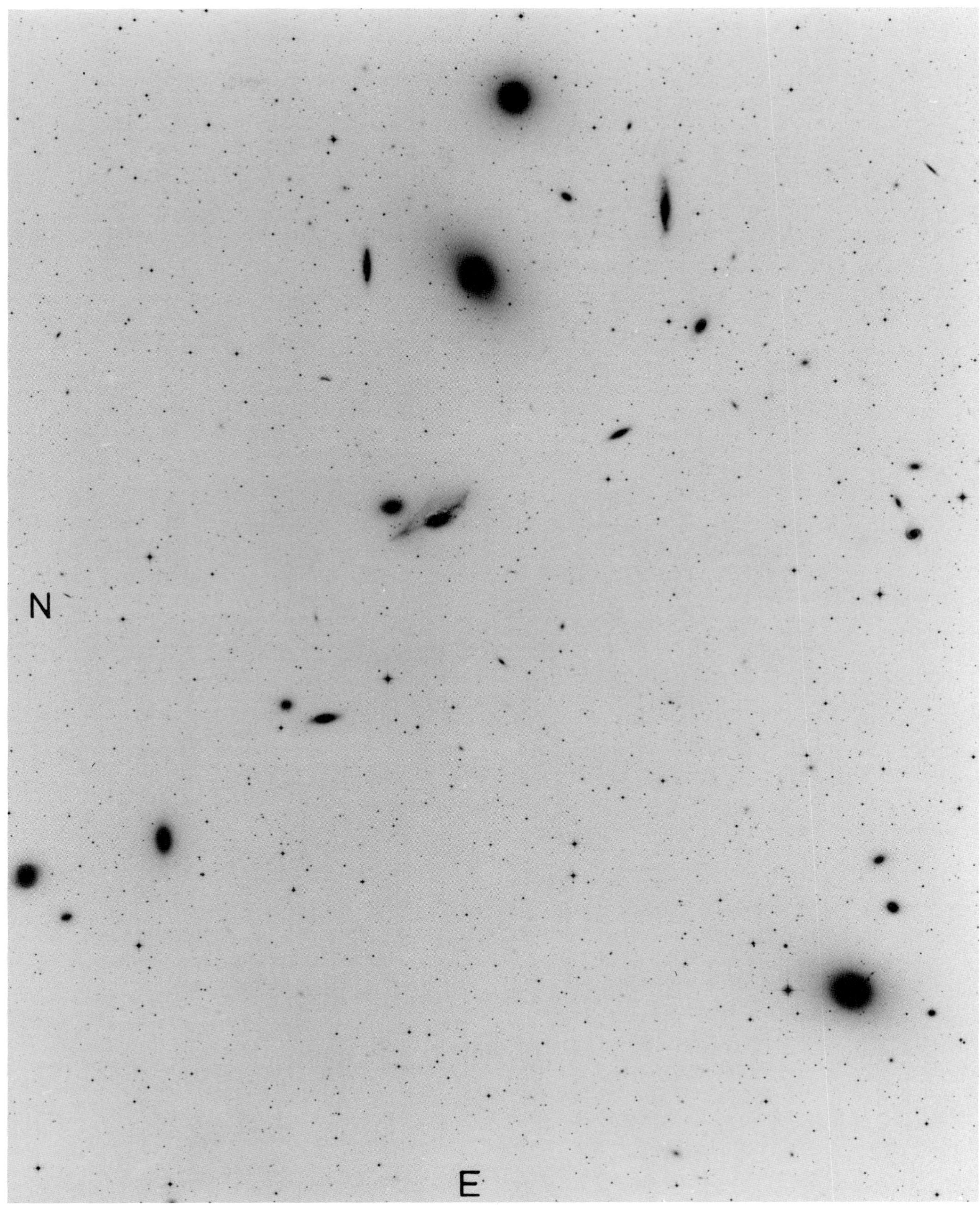

Plate 10.3

Plate 10.4

Plate 10.5

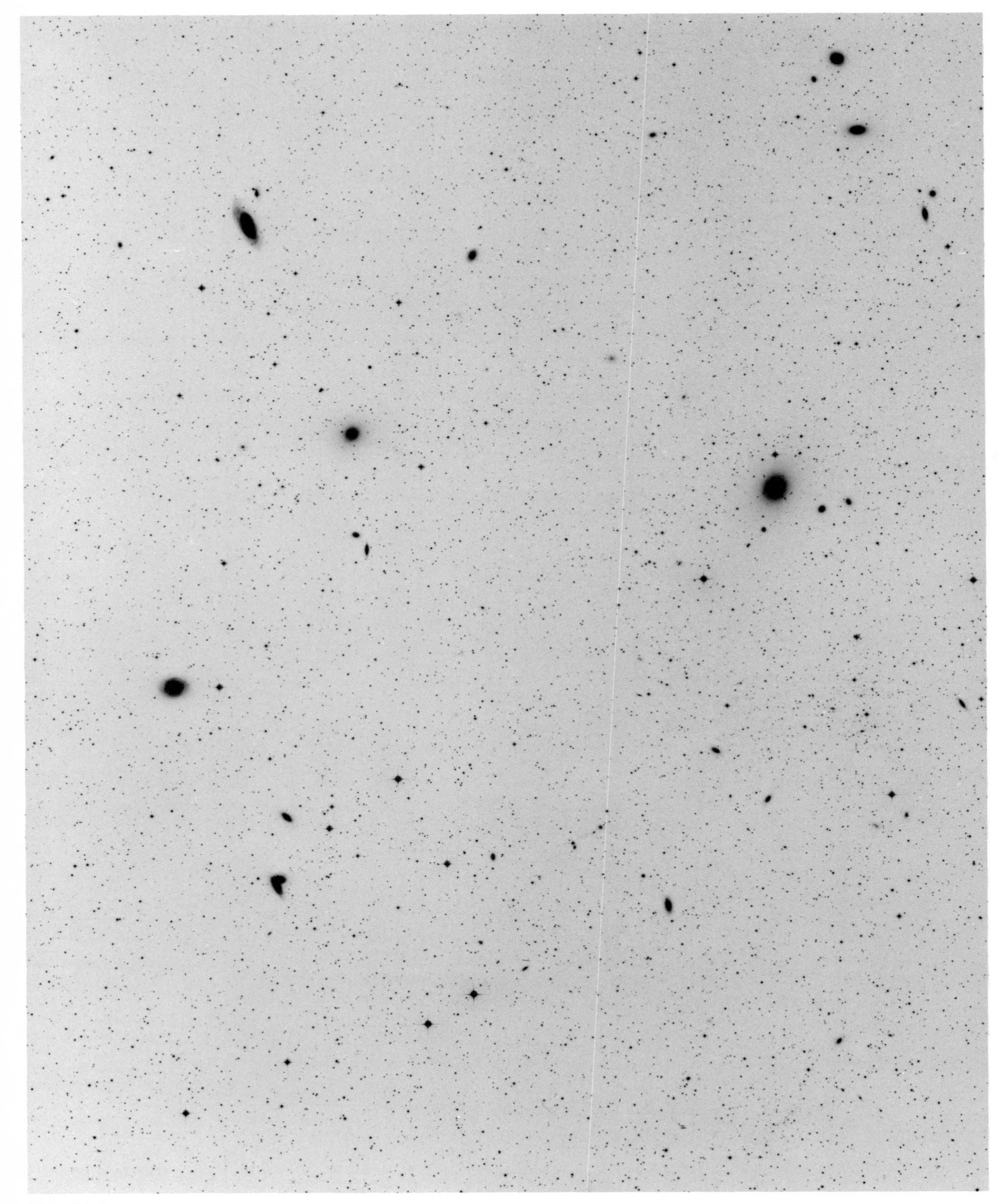

Plate 10.6

Plate 10.7

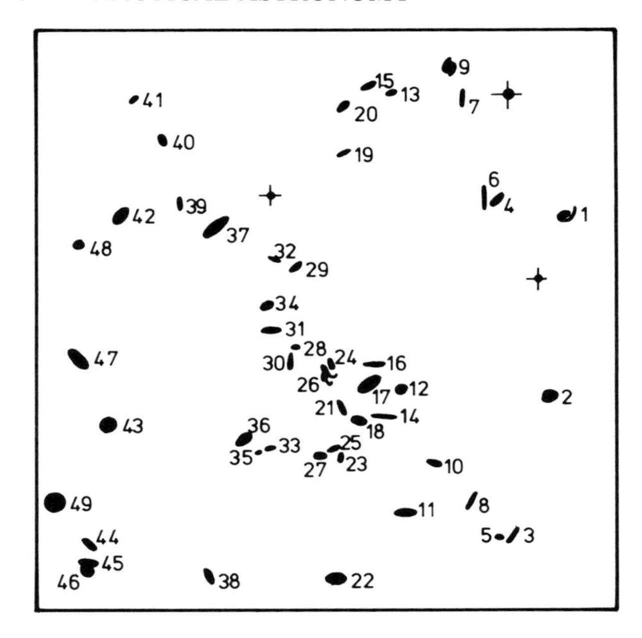

Figure 10.2 Identification chart of the galaxies in the Virgo cluster.

Suggestions. Large ellipticals of type E0 and large face-on spirals are the easiest to classify. Look carefully at objects which have an elongated shape to decide whether they are E types or spirals which are tilted to the plane of the sky. In both cases the centre of the image is dark, but the outer zones in the image of a spiral galaxy show irregularities whereas the outer zones of the elliptical galaxy are smooth. In the case of spiral galaxies it is more difficult to classify those which are tilted than those which happen to be viewed face-on. Points to take into account are that galaxies become flatter along the spiral sequence, so that those of type Sa show a central bulge when viewed edge on, and that spiral arms contain dust clouds which show up as dark streaks against the bright disk of a highly tilted galaxy. Even if you have difficulty in classifying a tilted spiral galaxy you may be able to decide if it has a bar. In some instances you may not be able to classify beyond the general 'S'.

SOLUTIONS AND DISCUSSION

1. The tables give the identification (NGC catalogue number), the Messier (M) catalogue number, if any, and the classification of the galaxies.

Table 10.1 Individual galaxies.

Plate 10.1		
a	NGC 1371	Sa
b	NGC 1232	Sc
c	NGC 1425	Sb
d	NGC 1433	SBa
e	NGC 4594	Sa
f	NGC 5101	SBa
Plate 10.2		
a	NGC 5058	Sc
b	NGC 5170	Sc
c	NGC 1300	SBb
d	NGC 3223	Sb
e	NGC 3717	Sb
f	NGC 1187	SBc
Galaxies on other pages		
Plate 4.1	NGC 253	Sb
Plate 11.1	NGC 224 (M31)	Sb
	NGC 221 (M32)	E2
	NGC 205 (M101)	E6
Plate 11.3	Large Magellanic Cloud	Irr
Plate 11.4	Small Magellanic Cloud	Irr

Table 10.2 Galaxies in the Virgo Cluster. (The numbers are those on the chart in figure 10.2. The classification Sab denotes galaxies which fall between classes Sa and Sb. Unusual galaxies, difficult to classify, are labelled 'pec' for peculiar. The galaxy NGC 4438 is distorted by the gravitational influence of its more massive neighbour.)

1	NCG 4254	Sc
2	NCG 4267	S0
3	NCG 4294	SBc
4	NCG 4298	Sc
5	NCG 4299	SB
6	NCG 4302	Sc
7	NCG 4312	Sab
8	NCG 4313	S
9	NCG 4321(M100)	Sc
10	NCG 4351	Sab
11	NCG 4371	SB
12	NCG 4374(M84)	E1
13	NCG 4379	S0
14	NCG 4388	Sb
15	NCG 4396	Sc
16	NCG 4402	Sb
17	NCG 4406(M86)	E3
18	NCG 4413	SBb(pec)
19	NCG 4419	SBa
20	NCG 4421	SBa
21	NCG 4425	SB(pec)
22	NCG 4429	S0
23	NCG 4431	S0
24	NCG 4435	SB
25	NCG 4436	S0
26	NCG 4438	Sb
27	NCG 4440	SBa
28	NCG 4458	E0
29	NCG 4459	S0
30	NCG 4461	Sa
31	NCG 4473	E5
32	NCG 4474	S0
33	NCG 4476	E5(pec)
34	NCG 4477	SBa
35	NCG 4478	E2
36	NCG 4486(M87)	E0
37	NCG 4501(M88)	Sb
38	NCG 4503	Sa
39	NCG 4516	Sab
40	NCG 4523	SB
41	NCG 4540	Sc
42	NCG 4548	SBb
43	NCG 4552(M89)	E0
44	NCG 4564	E6
45	NCG 4567	Sc
46	NCG 4568	Sc
47	NCG 4569(M90)	Sb
48	NCG 4571	Sc
49	NCG 4579	Sb

The smaller scale images of the galaxies of the Virgo cluster are more difficult to classify than the enlarged images of the central region. This illustrates the problem of classification of distant galaxies with very small images and the importance of having instruments which give the best possible resolution.

11 Nearby Galaxies

The Andromeda galaxy (M31, NGC 224)—or the Great Nebula in Andromeda, as it was called in earlier days—in the constellation of the same name at coordinates (0h 43m, +41°), is readily visible to the naked eye to observers in northern and low southern latitudes. With its hazy appearance and large size it has often in the past been mistaken for a comet. Its apparent brightness is the equivalent of a star of about fourth magnitude. Our Milky Way Galaxy and the Andromeda galaxy are the dominant members of a small but distinct group of 30 known galaxies which make up what is called the *Local Group*. The third largest member of the group is a spiral galaxy, M33, which is not far from the Andromeda galaxy in space. The other members are smaller galaxies many of which are associated with one of the two main members. Plate 11.1 shows the Andromeda galaxy and its closest satellites, the elliptical galaxies M32 (NGC 221) and M101 (NGC 205).

The Andromeda galaxy is a spiral galaxy of type Sb seen at an angle to the plane of the sky. It is an object which has been of immense importance in the progress of astronomy. Its outer regions were resolved into stars by Hubble in 1924 with the 100 inch (2.5 m) telescope at Mount Wilson and its core was resolved by Baade in 1944 with the 200 inch (5 m) telescope on Mount Palomar. The latter work led to the discovery of different stellar populations and established that M31 and other galaxies of the same type are broadly similar to our own Milky Way Galaxy. This advanced the study of the structure and constitution of our Milky Way and also the study of the Universe of galaxies on a larger scale. Baade also resolved stars in Andromeda's satellite galaxy, M32.

The distance to the Andromeda galaxy, 710 kpc, is known principally from observations of cepheid variable stars. Cepheid variable stars are pulsating stars which owe their regular changes in luminosity to expansion and contraction. The period of pulsation depends on the mean density of the star which in turn depends on its mass and size, quantities which also govern the star's luminosity. Thus the period of a cepheid variable star is an indication of its luminosity. Since the luminosities are very high—10^3 to 10^4 solar luminosities—these stars are recognisable and powerful beacons observable to large distances. The relation between period and magnitude for a set of such stars in the nearby Small Magellanic Cloud was established as far back as 1912 but it was not until the 1950s, after a great deal of difficult observational effort, that the distances of cepheids within our Galaxy and hence their absolute magnitudes were found. Cepheid variable stars are the best distance indicators in the realm of galaxies out to a distance of about 5 Mpc.

THE MASS OF THE ANDROMEDA GALAXY

The mass of our Galaxy (§5) is calculated from the period of rotation of its outer parts about its centre from Kepler's third law. The same method is used for finding the masses of external galaxies with observable rotation.

Spiral galaxies are particularly favourable because they have flat, practically circular disks and rotate, as our Galaxy does, in the plane of the disk.

A picture of how the mass is distributed within a galaxy may be built up by studying the pattern of speed of rotation as one moves outwards from the centre. To get the total mass of the galaxy it is necessary in principle only to know the rotational velocity at its extreme outer rim. Objects in those outer regions behave under the gravitational influence of the inner parts as if the whole mass of the galaxy were concentrated at one point at the centre. The motion of rotation is observed from Doppler shifts in the spectra of objects at different distances from the centre. Luminous blue stars and H II regions (containing ionised hydrogen, recognised from the emission line Hα in their spectra) are most commonly used for this purpose. These are very suitable also because they outline the spiral arms which occupy a thin plane within the general flat disk of the galaxy. Radio observations of H I (neutral hydrogen) in the arms also provide excellent Doppler patterns of galaxy rotation.

If a galaxy is seen edge-on, then one extremity shows maximum Doppler motion away from the observer and the other shows maximum Doppler motion towards the observer. If the galaxy is seen face-on, there is no motion in the line of sight. Most cases fall between these two extreme positions (figure 11.1(*a*)–(*c*)).

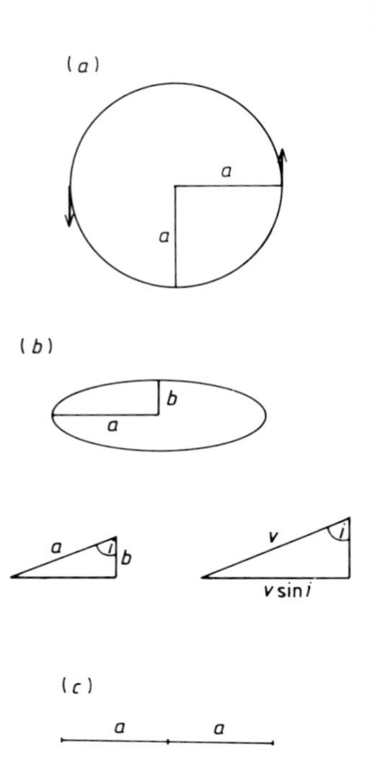

Figure 11.1 Aspects of a galaxy which has a flat circular disk of radius a. (*a*) Galaxy is in the plane of the sky and is seen face-on as a circle of radius a; rotational velocity is all in the plane of the sky and line of sight velocity is zero. (*b*) Galaxy is at an angle i to the plane of the sky; one axis is foreshortened to b where $b/a = \cos i$ and the observed line of sight velocity is $V \sin i$. (*c*) Galaxy at right angles to the plane of the sky appears as a straight line and the velocity is all in the line of sight.

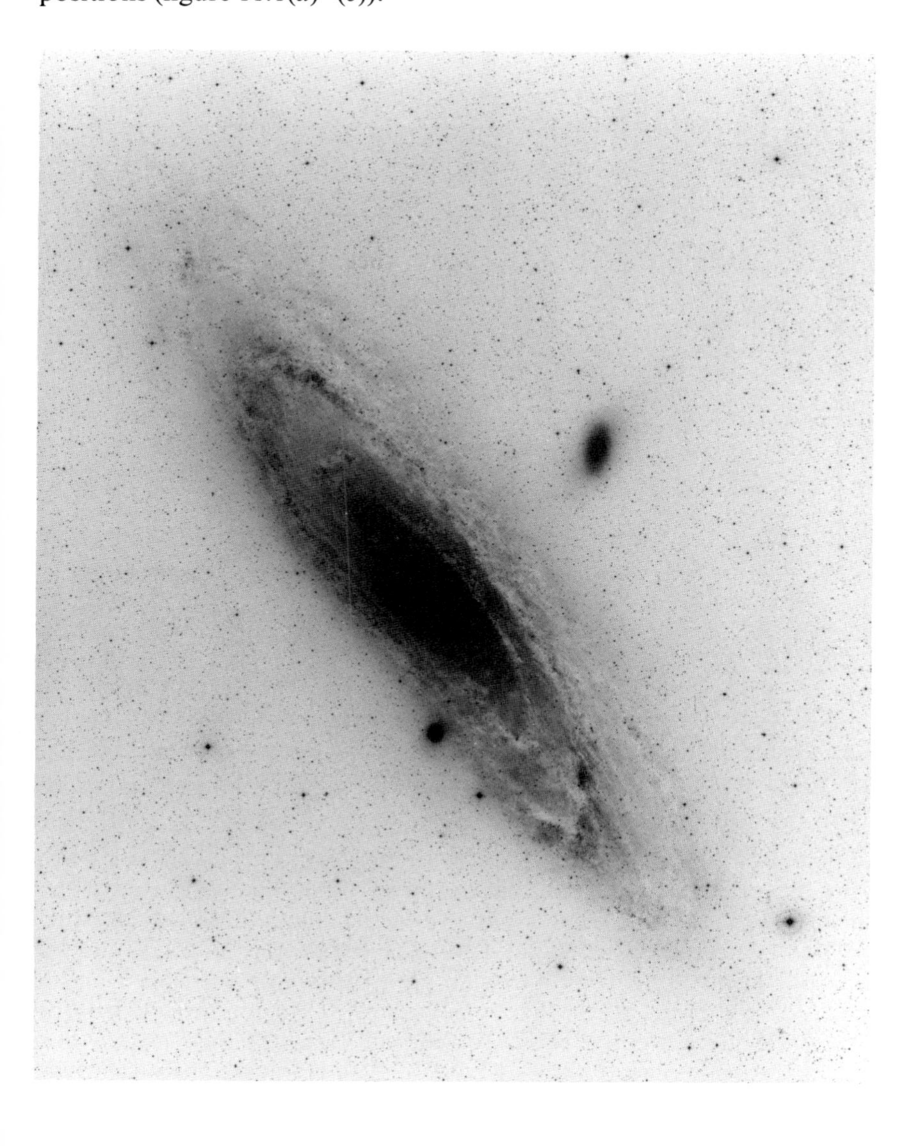

Plate 11.1 The Andromeda galaxy and two of its satellites. NGC 205 is the one on the right (© Palomar Observatory Sky Survey and Royal Observatory, Edinburgh).

A thin circular flat galaxy when viewed edge-on, appears to the observer as a thin line. The same galaxy, seen face-on, appears round. Examples of such edge-on and face-on galaxies are found in the sky and it is reasonable to assume that spiral galaxies in general may be described as approximately thin circular disks. (The halo contributes very little to the light of such a galaxy; the bright shape is almost entirely due to the disk.) If the galaxy is inclined at an angle i to the plane of the sky, the appearance of the disk is an ellipse. The diameter which is in the plane of the sky remains the same; the diameter at right angles to this is foreshortened. If a is the true diameter and b the foreshortened diameter their ratio is

$$b/a = \cos i.$$

If V is the velocity of rotation, the observed velocity in the line of sight is

$$V \sin i.$$

Figure 11.2 shows the line-of-sight velocity measured from Doppler shifts of objects at various points along the long axis of the Andromeda galaxy measured in arcminutes from the centre. Velocities are given with respect to the centre. Though one half of the galaxy is receding and the other is approaching, the graph shows the mean velocity of both sides without distinction of sign. Velocities in the inner regions, which are not needed for the calculations, are not plotted.

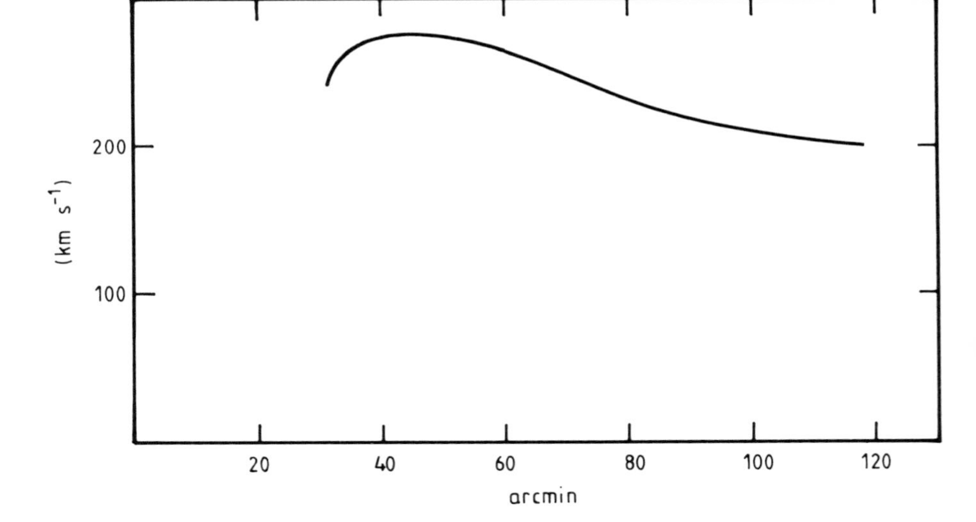

Figure 11.2 Rotation curve of the Andromeda galaxy. The velocity is the average of the velocities observed in the two halves (one of approach and one of recession) relative to the galaxy's centre without regard to sign.

Exercise 1. Calculate the mass of the Andromeda galaxy from its rotation curve, corrected for the effect of the tilt of the disk to the plane of the sky. The distance of the Andromeda galaxy is 710 kpc.

The calculation involves the velocity of rotation at the farthest point from the centre corrected for the inclination to the plane of the sky, and the distance of that point from the centre of the galaxy in linear measure. The angle of inclination is found from the shape of the galaxy in the photograph. The outline of an image at a fixed level of brightness is called an isophote; it is interesting to draw isophotes at two or three levels (by drawing on a transparent overlay) to see how constant their shapes are. Spiral galaxies have central bulges where the disk is thicker than elsewhere and where therefore the contours of the isophotes are less flattened than in the outer zones.

Suggestions. Measure the longest and shortest orthogonal dimensions of the outermost parts of the galaxy on the photograph in any convenient units. Calculate the angle i from the ratio b/a and look up sin i. Read off the value of v at the extremity of the rotation curve and divide by sin i to give the true velocity of rotation. Call this V. Measure carefully the angular distance of the extremity from the graph and convert to linear dimensions, using the known distance of the galaxy.

Two ways are suggested in which you can proceed.

(i) Use Kepler's third law (as in §5 for the mass of the Galaxy) which states:

$$M = R^3/P^2$$

where P is the period in years, R is the radius of the orbit in astronomical units and M is the mass of the system in units of solar mass. First convert R, the radius of the galaxy, to astronomical units, then to metres. The period of rotation is the circumference of the circle of this radius in metres divided by the velocity:

$$\text{Period (s)} = 2\pi R\,(\text{m})/V\,(\text{m s}^{-1}).$$

Convert the period to years. Now substitute the period (years) and the radius (AU) in Kepler's third law to get M, the mass of the Andromeda Galaxy in solar masses.

(ii) Alternatively you may use the formula for the velocity of an object in orbit of radius R metres around a mass of M kilograms. The formula is:

$$V(\text{m s}^{-1}) = (GM/R)^{1/2}$$

where G is the universal constant of gravitation. The result in this case is in kilograms. To convert to solar masses, divide by the solar mass in kilograms. (See Appendix 1 for conversions and other data.)

THE MAGELLANIC CLOUDS

The Magellanic Clouds, the nearest of all galaxies and the principal satellites of the Milky Way Galaxy, are visible to the naked eye in the southern hemisphere. The larger and brighter is at coordinates (05h 21m, −69°) and the smaller at (00h 30m, −75°). With their very high southern declinations the Clouds are conspicuous to observers in southern latitudes and never set at latitudes above 20°S. The Large Magellanic Cloud (LMC) is bright enough—with a total brightness equivalent to a star of magnitude 0—to be seen even at full moon. The Small Magellanic Cloud (SMC) is equivalent in total brightness to a star of magnitude 2.4. The Clouds were named after the Portuguese navigator Magellan by his companion Pigafetta in 1521, who was the first to record them, though they were well known before then to sailors plying the African coast. The word 'clouds' aptly describes their appearance to the naked eye, but when viewed through a telescope they resolve into numerous stars and nebulosities, like the Milky Way (plate 11.2).

Plate 11.3 is a photograph of the Large Magellanic Cloud in the U (near-ultraviolet) waveband. The area of the photograph (6.5° × 6.5°) is not wide enough to cover the entire cloud, which extends a few degrees further in all directions. On this photograph the elongated bar and the outlying regions of bright nebulosities which resemble untidy spiral arms are clearly seen; in the galaxy classification scheme (§10), the LMC falls logically beyond type Sc, the loosest type of spiral; it is sometimes classified as lying between Sc and Irr. The bright nebulosities are H II regions (emitters of ultraviolet radiation in certain wavelengths) and supernova remnants. The largest H II region is 30 Doradus, mentioned in connection with Supernova 1987A (§9). You will observe the shell-like structure of some of the nebulosities and may find it interesting to compare their shapes and sizes with the supernova remnants of our own Galaxy (§9).

Plate 11.4 shows the field of the Small Magellanic Cloud in the J (blue) waveband. It includes two conspicuous globular clusters belonging to our own Galaxy, NGC 104 and NGC 362, discussed earlier (§7), which are in the foreground and unconnected with the Cloud. On this photograph the SMC appears more populous than the LMC on the previous plate. This is the effect of the use of different wavebands in the photographs: the very luminous hot blue stars of the LMC are recorded in the U band but not the more numerous cooler stars which are too faint to show up at the distance of the LMC. These stars are included in the broader waveband used for the SMC photograph.

The distances of the LMC and the SMC are 50 kpc and 63 kpc respectively. Cepheid variable stars and other luminous stars recognised from their spectra are available as distance indicators in the Magellanic Clouds. The variable RR Lyrae stars which are found in globular clusters and also individually in the Galaxy are also there. These stars are particularly reliable distance indicators. Observed in the Magellanic Clouds, they tie the distances of the Clouds to local distances within the Galaxy. The more luminous cepheids in turn tie the distances of the Clouds to the distances of other galaxies. In this way the Magellanic Clouds play an important role in building up the distance scale of the universe.

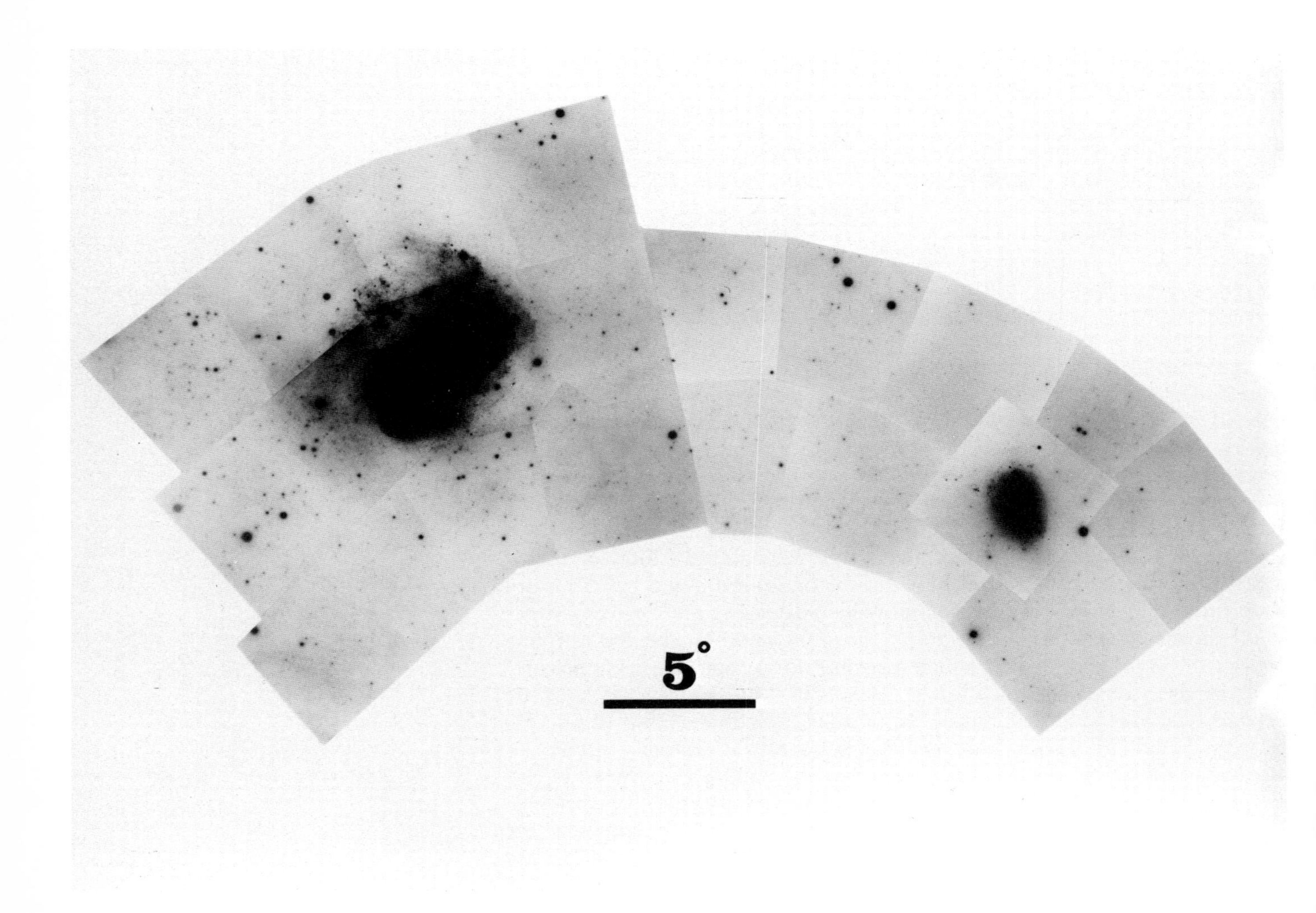

Plate 11.2 The Large and Small Magellanic Clouds; a composite photograph made up of several fields observed with the UK Schmidt Telescope (© Royal Observatory, Edinburgh).

Plate 11.2 shows the two Magellanic Clouds on a composite photograph made up of a number of photographs taken in the blue waveband with the UK Schmidt Telescope. A point of interest in the composite photograph is the way in which the individual fields overlap and converge towards the bottom of the photograph as they point to the south pole of the sky. The photograph (on which the inner parts of the galaxies are overexposed) records very faint stars in the outer regions, and shows the considerably wider extent of the LMC than is seen on plate 11.3. Radio astronomical observations reveal that both Clouds are immersed in a huge envelope of neutral hydrogen gas. It is worth calculating the linear separation of the two Clouds to establish whether they are indeed close together in space.

Exercise 2. From plate 11.2 find the separation in space of the Magellanic Clouds. The Large and Small Magellanic Clouds are at distances 50 kpc and 63 kpc respectively.

Plate 11.3 The Large Magellanic Cloud in the U waveband (© Royal Observatory, Edinburgh).

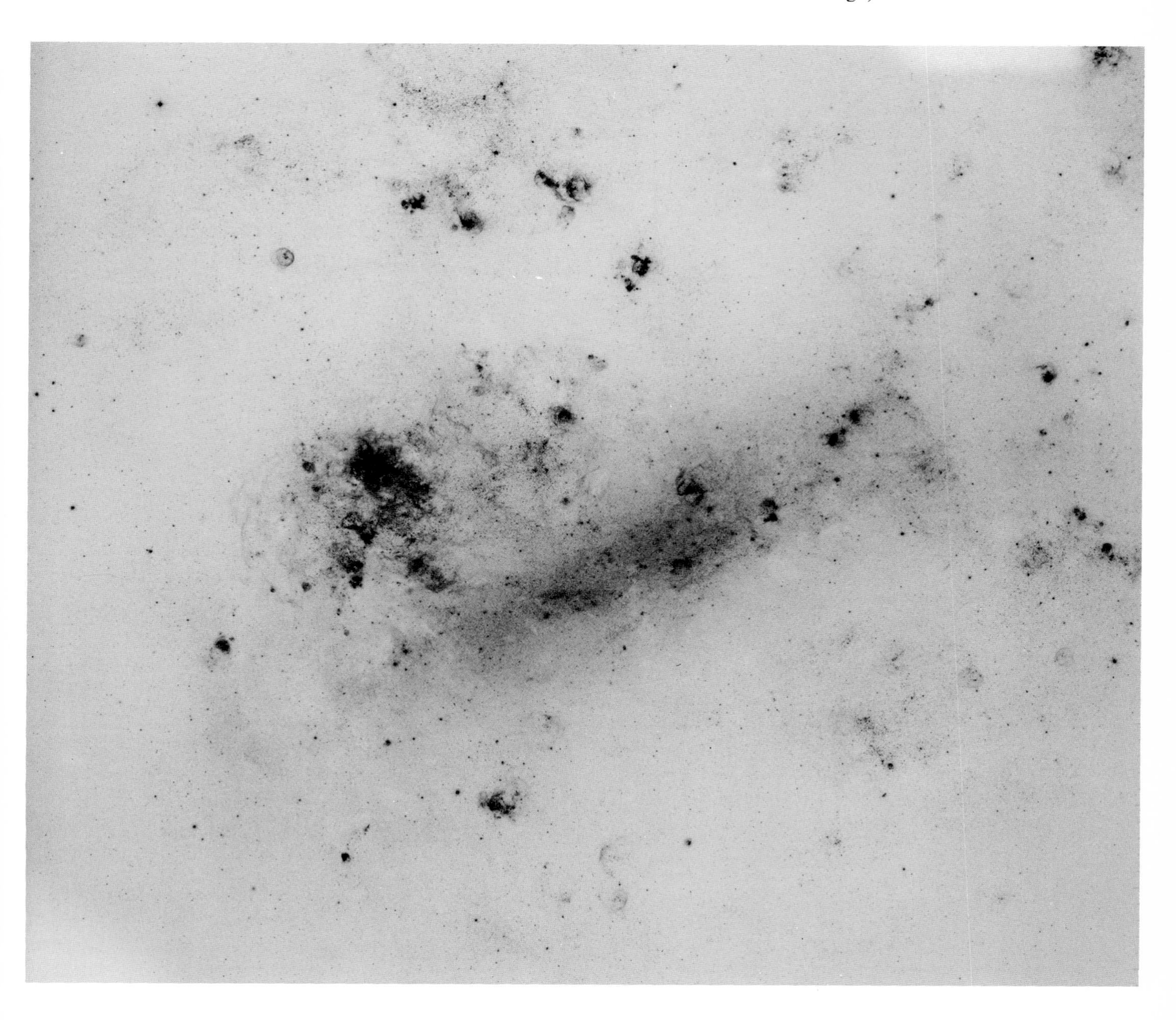

Strictly, the distance of widely separated objects on the sky should be measured on a globe. In this case, however, the projection of part of the sky on the composite photograph is adequate for our purpose.

Suggestions. Try to fix the centre of each Cloud on the composite photograph. Measure the separation of the centres with a ruler and, making use of the scale marked on the photograph, convert this to degrees. Construct to scale on graph paper a triangle showing the Sun (our position) and the two Clouds. The side of the triangle joining the Clouds gives the separation of the Clouds in space. The triangle may also be solved by trigonometry.

THE MASS OF THE SMALL MAGELLANIC CLOUD

The masses of the Magellanic Clouds are found from their rotation curves by the method described in Exercise 1, though on account of their irregular

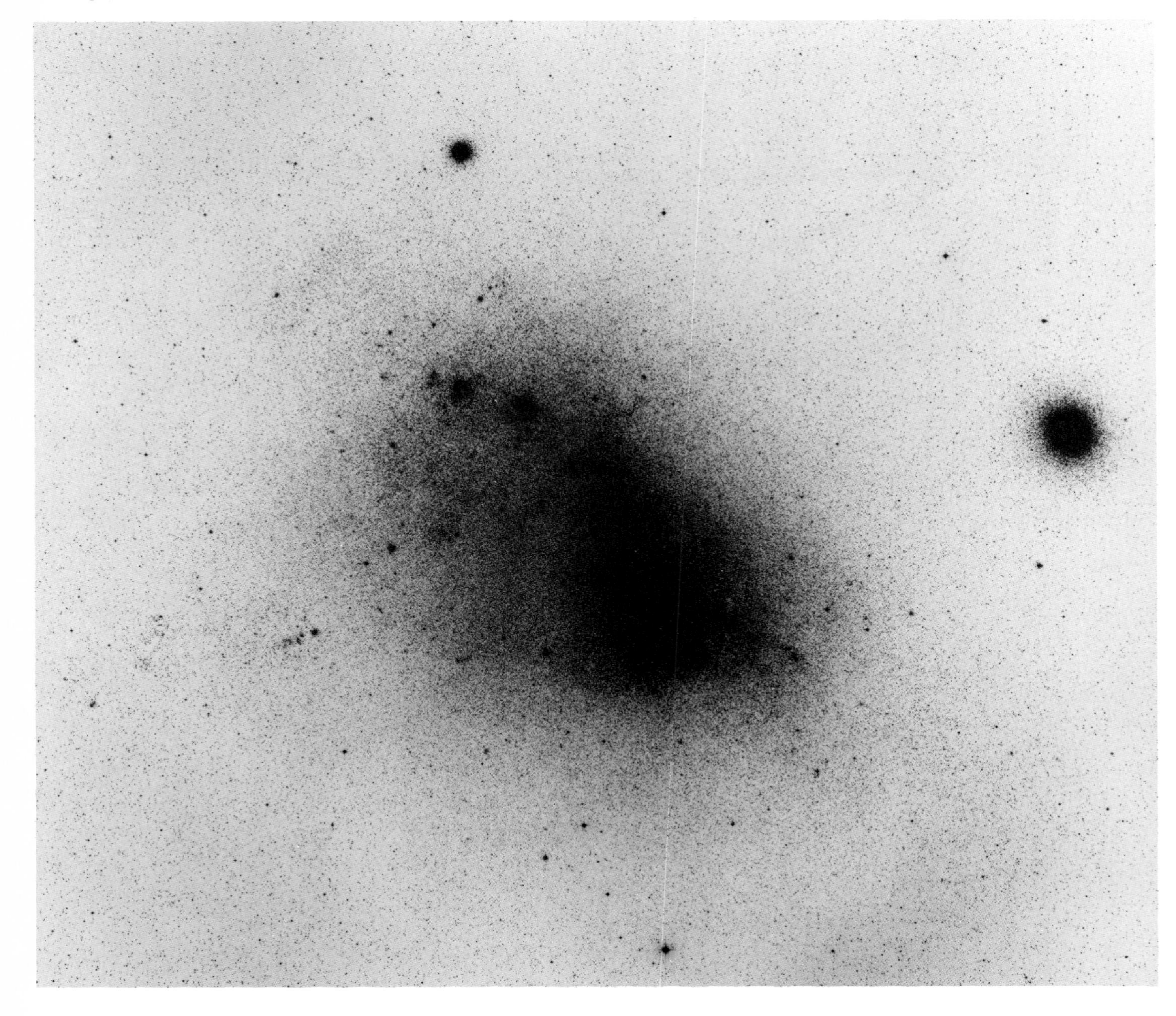

Plate 11.4 The Small Magellanic Cloud. The two globular clusters are in the foreground: the larger is 47 Tucanae (NGC 104) and the smaller is NGC 362 (© Royal Observatory, Edinburgh).

shapes the planes of their disks are less clear-cut than in the case of a regular spiral galaxy like the Andromeda galaxy. Figure 11.3 shows the SMC's rotation curve recorded with a radio telescope tuned to 21 cm and adjacent wavelengths. Radiation at the wavelength 21 cm is emitted by neutral hydrogen (H I) in the interstellar medium. Motions in the gas are recognised by the Doppler shift, just as in optical observations. The Magellanic Clouds have a high content of neutral hydrogen gas, which delineate their disks. The pattern of motion of neutral hydrogen in the SMC is complex and difficult to unravel. One strong feature, however, is a maximum gradient of radial velocity across the galaxy in the direction from south west to north east. This is interpreted as demonstrating the rotation of the Cloud around an axis at right angles to this direction.

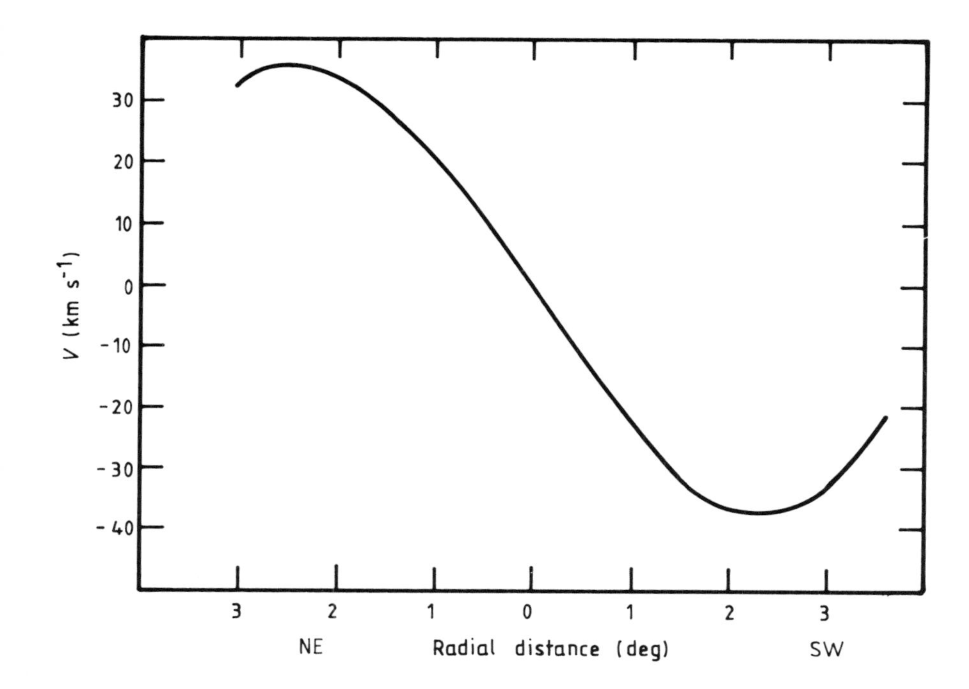

Figure 11.3 Rotation curve of the Small Magellanic Cloud from 21 cm radio observations.

Exercise 3. Find the mass of the Small Magellanic Cloud from its radio rotation curve recorded in the direction from south west to north east. The outline of its shape on plates 11.2 and 11.4 is assumed to represent a flat tilted circular disk.

SOLUTIONS AND DISCUSSION

1. The inner isophote is less elongated than the outer ones and demonstrates the bulge as expected. For the outer isophotes the mean ratio of the shortest to the longest axis is $0.21 = \cos 78°$.

$v = V \sin 78° = 0.98\ V$
v from curve $= 200 \times 10^3\ \mathrm{m\,s^{-1}}$
$\therefore V = 204 \times 10^3\ \mathrm{m\,s^{-1}}$.

The distance to the galaxy $r = 710$ kpc $= 2.2 \times 10^{22}$ m. At the extremity $R = 118$ arcmin $= 3.4 \times 10^{-2}$ rad

$\therefore$ radius of galaxy $= 3.4 \times 10^{-2} \times 710 = 24$ kpc $= 7.5 \times 10^{20}$ m $= 5.0 \times 10^9$ AU.

The period of rotation $= 2\pi R/V$ (R in metres, V in $\mathrm{m\,s^{-1}}$)

$= 23 \times 10^{15}$ s $= 7.3 \times 10^8$ years.

By Kepler's third law $M = R^3/P^2 = 2.3 \times 10^{11}$ solar masses.

Alternatively, from the velocity formula, $M = V^2R/G$

$$M = (2.04 \times 10^5)^2 \times 7.5 \times 10^{20}/6.7 \times 10^{-11} = 4.6 \times 10^{41}\ \text{kg} = 2.3 \times 10^{11}\ \text{solar masses.}$$

These results, for the angle of inclination and the mass, agree with those adopted by astronomers from a complete analysis of the rotation curve.

The calculation assumes that all the mass of the galaxy is contained within a radius of 24 kpc, the visible limit. In fact radio observations of H I (neutral hydrogen gas) trace the Andromeda galaxy to 150 arcmin (30 kpc) from the centre along the major axis (figure 11.4).

According to the velocity formula for an object in orbit around a mass M (V m s^{-1} = $(GM/R)^{1/2}$), if M remains the same, V falls off inversely as the square root of R. If the observations have truly reached the extremity of the galaxy, the last part of the rotation curve ought to fall off inversely as the square root of the distance. Such rotation is called 'Keplerian' because it obeys Kepler's laws. Figure 11.4 shows that this is not the case in the present instance. Optical and radio observations of many spiral galaxies tend to record rotation curves which do not fall off as expected.

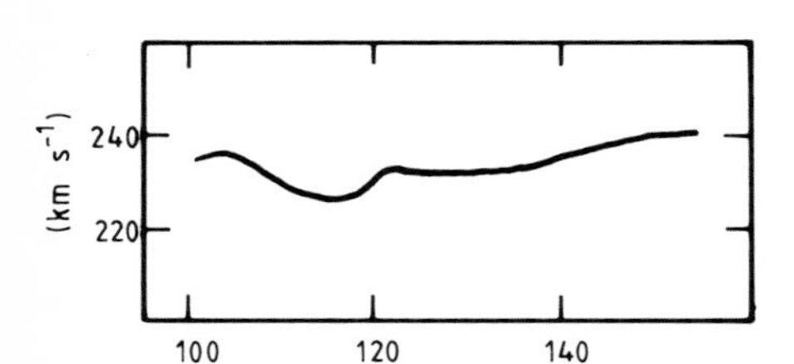

Figure 11.4 Radio rotation curve of the Andromeda galaxy which extends further than the optical curve and does not fall off in a Keplerian fashion.

Analysis of the rotation curve of the Andromeda galaxy out to 30 kpc which reconstructs the variation in density as one goes outwards, gives 2.5×10^{11} solar masses as the mass of the Andromeda galaxy to that limit. For the entire galaxy including that beyond this limit, a mass of 3×10^{11} solar masses is generally adopted.

The distance to the Andromeda galaxy, according to some observers, may be larger (900 kpc) than the value which we have used. The linear radius converted from the angular radius scales directly with the distance. The mass is directly proportional to the radius (in the velocity formula). Therefore the mass scales directly with the distance, and the larger value of 900 kpc would make the mass greater by a factor of 900/710, raising the mass to 3.8×10^{11} solar masses.

It is worth noting that in this particular example the correction for inclination makes very little difference to the result because the galaxy is so nearly at right angles to the plane of the sky.

2. The triangle to be solved is similar to that used in §3. The angle between the Clouds is 22° and their linear separation is 25 kpc. This is less than the diameter of the Galaxy and about half of the distance of the Clouds from the Galaxy. The Clouds may therefore be regarded as a pair, rather like a double star.

3. The ratio of the long and short axes of the Small Magellanic Cloud is 0.5; the inclination of the disk to the plane of the sky is therefore 60°.

Maximum radius (mean of two sides) = 3.3° = 7.5×10^8 AU,
velocity at the extremity (mean of two sides) = 28×10^3 m s^{-1}/sin 60°,
$P = 6.9 \times 10^8$ years,
$M = R^3/P^2 = 9 \times 10^8$ solar masses.

Unlike the calculation in Exercise 1, the correction for inclination is significant in this example. The adopted mass is 10^9 solar masses. The interesting feature of the result is the great difference between the mass of the SMC and that of a full-sized galaxy like the Andromeda or Milky Way galaxies. It is an important result because it shows that, though small galaxies may be fairly common, they contribute very little to the mass of the universe of galaxies as a whole.

Table 11.1 The Local Group of galaxies.

		r (kpc)	Type	Mass	M_V
1	Andromeda Galaxy (M31)	710	Sb	3×10^{11}	−21.1
2	Milky Way Galaxy	8	Sb	2×10^{11}	−20.5
3	Triangulum Galaxy (M33)	730	Sc	1×10^{10}	−18.9
4	Large Magellanic Cloud	50	Irr	1×10^{10}	−18.5
5	NGC 205	710	E6	1×10^{10}	−16.4
6	IC 10	1260	Irr	4×10^9	−17.6
7	M32	710	E2	3×10^9	−16.4
8	Small Magellanic Cloud	63	Irr	1×10^9	−16.8
9	NGC 185	710	dE0	1×10^9	−15.2
10	NGC 147	710	dE4	1×10^9	−14.8
11	NGC 6822	470	Irr	4×10^8	−15.7
12	NGC 1613	740	Irr	3×10^8	−14.8

The distances, masses, galaxy types and absolute magnitudes of the first 12 members (by mass) of the Local Group are listed for interest in table 11.1, from which it is seen that the Andromeda and Milky Way galaxies make up over 90% of the total mass of the group. The centre of mass of the group lies therefore at a point in space between these two large galaxies, but nearer to the Andromeda galaxy, because it is the more massive. Four of those listed are companions of the Andromeda galaxy; two are companions of our Galaxy. Both large galaxies have other smaller companions, the Galaxy having seven dwarf spheroidal satellite galaxies with masses below 10^8 solar masses. Note that the absolute magnitudes of the galaxies do not keep step with their masses. The reason is that young luminous stars make a contribution to the luminosity of a galaxy which is vastly greater than their contribution to the mass. It is an interesting exercise to calculate the ratio of mass to luminosity for the various members of the Local Group. (Absolute magnitude may be converted to solar luminosities as in Exercise 2, §9.) The ratio, M/L, is a useful and widely used index of stellar population type. An index of 1 represents a population consisting on average of stars like the Sun; a lower index signifies a higher proportion of luminous stars.

12 Clusters of Galaxies

The further one looks into space, the more galaxies one observes. Their overall numbers increase with magnitude up to great distances at the rate expected for a space which is uniformly filled with them. An exception is the plane of the Galaxy and especially the southern Milky Way where interstellar dust hides from view the galaxies beyond, creating what used to be called the 'zone of avoidance'.

The Virgo cluster of galaxies and the Local Group of galaxies have been discussed earlier (§10 and §11). These groups are not just local irregularities in the universe of galaxies. When the distribution of galaxies is studied it emerges that they have a general tendency to cluster. About 50 clusters have been counted within 20 Mpc of our Galaxy. Clusters vary in size and membership from small ones like the Local Group to huge clusters like the one in Virgo which has thousands of galaxies and dominates the local region of space. The clusters themselves often belong to even larger groupings; there are several other clusters in the vicinity of the large Virgo cluster.

Clusters of galaxies, like individual galaxies, lend themselves to classification. There are regular symmetrical clusters in which galaxies concentrate on one or two giant elliptical galaxies at the centre surrounded by outlying mainly spiral galaxies, and irregular ones which are more loose and may have more than one concentration. Clusters are also described as rich or poor according to the numbers of their members. The Local Group and the Virgo cluster are both irregular, one poor and one rich. Some clusters, like the Local Group, are devoid of giant elliptical galaxies; all clusters contain spiral galaxies.

The classification of clusters is useful for the study of the origin and dynamics of the clusters and their members: there must be a reason why clusters evolved in these fairly distinct ways, for example, giant elliptical galaxies in cluster centres may be the result of mergers of two or more galaxies. Classification is also useful as an aid to estimating distances where other methods are difficult. A very distant cluster, in which individual members cannot be accurately classified, may be recognisable by its overall structure and may be assumed to be similar in intrinsic properties to other clusters of its class.

The Virgo cluster of galaxies is shown in plates 10.3–10.7. Photographs taken with the UK Schmidt Telescope of three more clusters of galaxies and the central part of one of them are shown on plates 12.1–12.4. The clusters are named after the constellation in which they are found, or by a catalogue number.

Exercise 1. Classify as regular or irregular the clusters of galaxies on plates 12.1–12.3 and confirm the classification of the Virgo cluster as irregular.

Suggestions. Though the answers may appear to strike the eye immediately, impressions can be misleading. Star images may confuse the picture, and it is easy to be biased by the fortuitous centering of the photograph. Make a map for each cluster by tracing the images of the galaxies on a sheet of clear overlay. Mark the elliptical and spiral galaxies in different

Facing page:

Plate 12.1 The Centaurus II cluster of galaxies (© Royal Observatory, Edinburgh).

Page 94:

Plate 12.2 The Pavo cluster of galaxies (© Royal Observatory, Edinburgh).

Page 95:

Plate 12.3 The Hydra I (Abell 1060) cluster of galaxies (© Royal Observatory, Edinburgh).

Plate 12.1

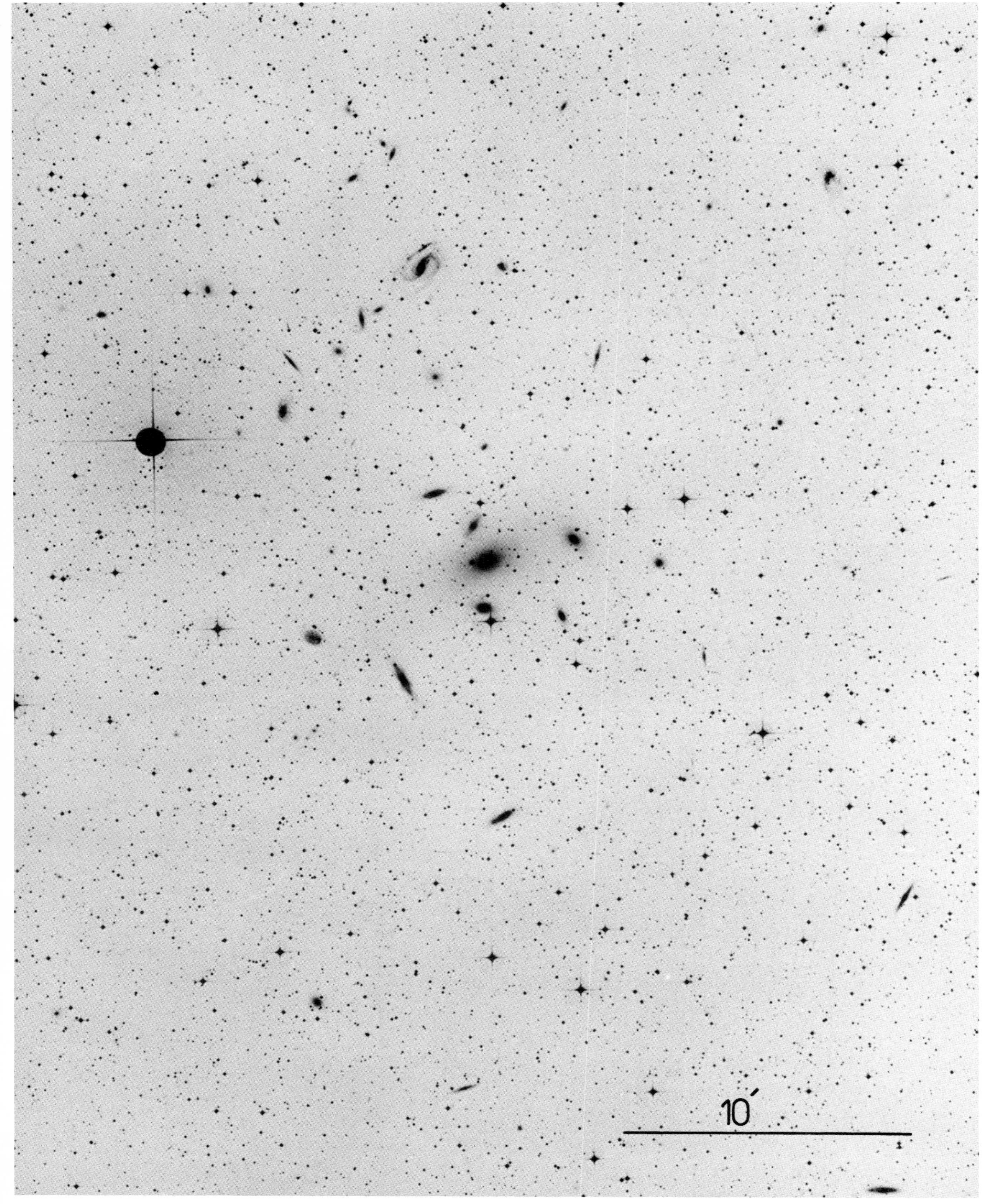

Plate 12.2

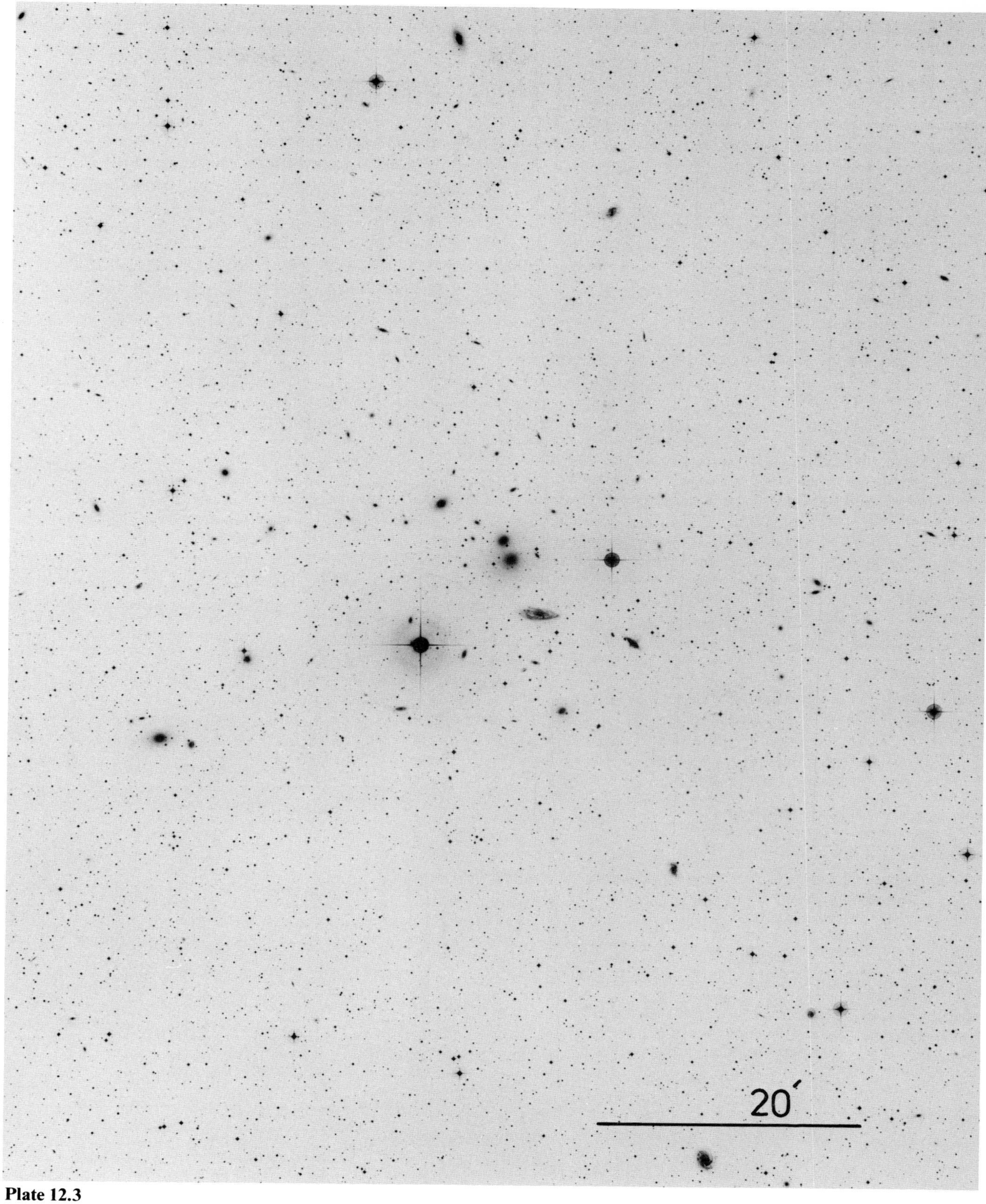

Plate 12.3

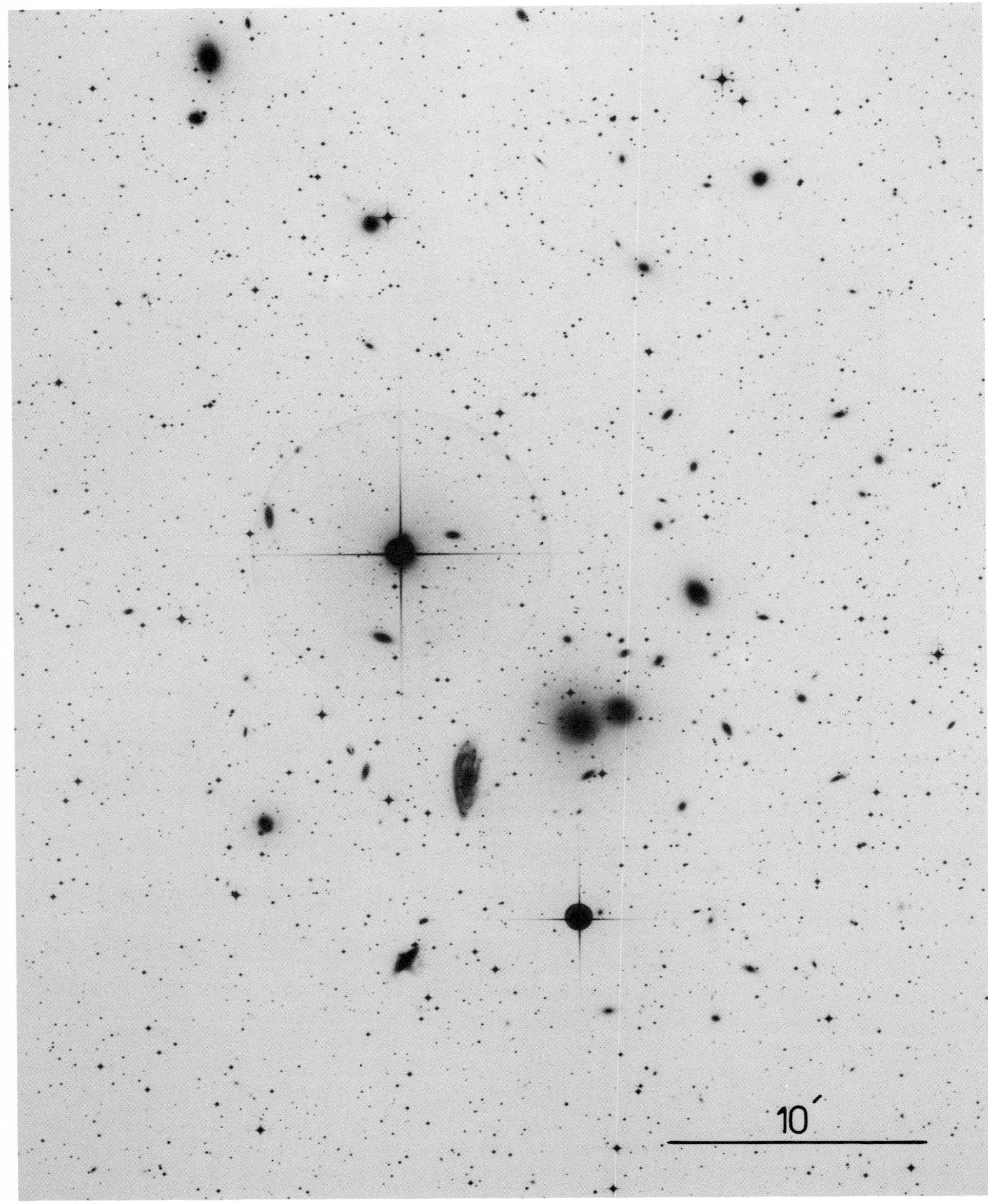

Plate 12.4

colours. Try as well as you can to classify the smaller images correctly. Images of stars above a certain limit of brightness show spikes; images without spikes down to this limiting size must therefore be galaxies. Ignore images below this threshold which cannot be separated into stars and galaxies by eye. It is not critical if you cannot decide whether particular small galaxies are spirals or ellipticals; you may like, for the sake of thoroughness, to mark the unclassified galaxies in a third colour. There are, of course, galaxies other than the cluster members present; these are foreground or (more likely) background galaxies scattered uniformly throughout the field.

When you remove the overlay and place it over a sheet of white paper, you will gain an unbiased impression of the distribution of galaxies in the field. To help with classification, draw a series of concentric circles on the sheet of white paper and move the overlay map about until you find the centre of the cluster. The centre will not necessarily fall on one particular galaxy. In the case of an irregular cluster, the point of greatest density will not be in the centre. Its centre is the position in which the galaxies are as symmetrically placed as possible in two perpendicular directions with respect to the centre of the circles.

Exercise 2. Estimate the distance of the cluster Abell 1060 by comparing the brightest members (plate 12.4) with the corresponding members of the Virgo cluster, which is at a distance of 20 Mpc.

One of the methods of measuring the distance of an extended object is from its apparent angular diameter. To apply the method to galaxies in clusters, it is assumed that the brightest members in a distant cluster are similar to their counterparts in a nearby cluster of the same type.

From the previous exercise it will have been found that Abell 1060 is an irregular cluster, similar in type to the Virgo cluster. The distance to the relatively nearby Virgo cluster is known with confidence from observations of many of its individual members.

Suggestions. Choose for this exercise the five first ranking galaxies in the cluster Abell 1060. Measure on the photograph, to the best accuracy you can, the diameters of the five largest galaxies in the cluster, noting whether each one is a spiral or an elliptical. Do the same for the first ranking galaxies in the Virgo cluster (the photographs of the clusters, though not covering the entire cluster area, include all the brightest members). Photographs of both clusters may be assumed to record the light of the galaxies to the same isophote.

Convert the measurements to arcminutes. Take the ratios of the Virgo galaxy diameters to those in the other cluster, comparing ellipticals with ellipticals and spirals with spirals, i.e., the largest elliptical with the largest elliptical; the largest spiral with the largest spiral, the second largest elliptical with the second largest elliptical etc. The five ratios may thus be made up of two ellipticals and three spirals, or three ellipticals and two spirals, or some other combination.

The mean ratio of the dimensions of the corresponding galaxies in the two clusters is the ratio of their distances. Multiply this ratio by 20 Mpc (the distance of the Virgo cluster) to give the distance of the cluster. The deviations from the mean ratio will also give you the probable error in the distance.

Exercise 3. The entire extent of the Pavo cluster is covered by the photograph in plate 12.2. Estimate the number of galaxies in this cluster. Given that its distance is 100 Mpc, compare the Pavo cluster with the Local Group as regards dimensions and number of members.

Suggestions. With the aid of the grid of concentric circles in Exercise 1 it is possible to assess the dimensions of the cluster. If the cluster is not symmetrical, mark the radius at which the galaxy density merges into the background in more than one direction and outline the cluster with an elliptical contour.

It is not satisfactory to attempt to guess which galaxies belong to the cluster and which belong to the general field. It is better to estimate how many galaxies belong to the general field and to subtract this number from the total number counted. To do this, mark off some rectangular areas in easily measurable units (say 2 cm × 2 cm) well outside the contours of the cluster on your galaxy map and count the number of galaxies within those areas. Calculate

Facing page:

Plate 12.4 The centre of the cluster Abell 1060 (© Royal Observatory, Edinburgh).

the number per square and multiply by the area of the whole field in the same units. Subtract this number from the total number of galaxies to get the number in the cluster.

Another way is to return to the actual photograph and to count *all* images in the field, both stars and galaxies, down to a chosen limit. Divide the whole field into squares of a convenient size, and count the images one square at a time. Add these up to get the total number of images in the field. Now mark off, as already described, some sample areas which are well outside the contour of the cluster. Calculate the mean number per square and multiply by the area of the whole field in the same units. Subtract this number from the total number of images in the field to get the number of cluster members. It is assumed that both stars and background galaxies are uniformly distributed over the field and that therefore the residual count represents cluster galaxies only. The advantage of this procedure is that you avoid the problem of distinguishing galaxies from stars among the fainter images.

THE TOTAL MASS OF A CLUSTER OF GALAXIES

In §11 the question of the masses of individual galaxies was discussed. For a galaxy which has a rotating disk the mass may be deduced directly from its rotation curve. The method is not possible, however, in the case of elliptical galaxies which have no particular plane of rotation. Instead, an indirect method is used which depends on knowing the balance of energy within the galaxy. The stars which make up the system have kinetic energy from their individual motions within it; they also have potential gravitational energy. The internal motions of the stars reveal themselves through the Doppler effect in the galaxy's spectrum. The potential energy regulates the dimensions of the galaxy. Globular star clusters can be examined by this method, as mentioned in §7. So also can clusters of galaxies. The method is an application of what is called the 'virial theorem', from the Latin 'vires' meaning force or energy.

The virial theorem states that in a system which is neither expanding nor contracting, the gravitational potential energy is numerically twice the kinetic energy. The kinetic energy comes from the motions of the individual objects within the system which is held together by gravity. If the kinetic energy is too high, gravity is not able to hold it together and the objects fly away. If the converse is the case, the objects fall in towards the centre and the system contracts.

The principle is particularly easy to illustrate in the case of galaxies in clusters, and in the exercise which follows the mass of a cluster of galaxies is derived from observed data of its dimensions and internal motions.

Assume an average mass m for each galaxy, and a total of N galaxies in the cluster. A galaxy moving inside the cluster with velocity v has kinetic energy $\frac{1}{2}mv^2$. The individual galaxies have different velocities but we take this to be the average kinetic energy. The total kinetic energy of the N galaxies is therefore $\frac{1}{2}mNv^2$.

A galaxy moves in three dimensions. The velocity in one of these dimensions, the line of sight, is the radial velocity, which is observed from the Doppler shift in the galaxy's spectrum. The velocities in the other two dimensions which are in the plane of the sky (proper motion) are unobservable (because the galaxies are too distant). If the three components of velocity are U, V and W, the combined velocity v is the square root of the sum of their squares, or $v^2 = U^2 + V^2 + W^2$. If there are several objects and no preferential direction, the averages of U^2, V^2 and W^2 are the same and equal to $\frac{1}{3}v^2$. This means that if V^2 is the average of the squared observed radial velocities, the total kinetic energy of the cluster is $\frac{3}{2}NmV^2$.

The potential energy has now to be calculated. The potential energy of two objects each of mass m separated by a distance r is Gm^2/r where G is the constant of gravitation.

There are N galaxies in the cluster, each possible pair among which has potential energy according to that formula. The total potential energy is the sum of these amounts, added up for all possible pairs. This turns out to be GN^2m^2/R, where R is called the 'effective radius' of the cluster which is of the order of the actual geometrical radius. Its precise value may be found by an analysis of the distribution of objects as seen in projection on the sky; this analysis is too lengthy to be included here.

The potential energy is twice the kinetic energy, so

$$GN^2m^2/R = 3NmV^2.$$

Simplifying this equation gives the total mass M of the cluster (N times the individual mass m of a galaxy):

$$M = Nm = 3V^2R/G.$$

In the exercise the mass of the cluster Abell 1060 is derived from observed data of its dimensions and internal motions, which are accurately known. The cluster is shown on plate 12.3; however, the photograph does not extend far enough to show the cluster in its entirety and we therefore accept the effective radius and the total number of galaxies from other sources. The radial velocities are observed spectroscopically from the Doppler shift and are listed for a sample of members in table 12.1. (It is unnecessary to identify these galaxies individually on the photograph.) The radial velocity in the case of each galaxy consists of two elements—the velocity of recession of the galaxy on account of the expansion of the universe, and the local velocity of the galaxy as a member of the cluster (figure 12.1).

Exercise 4. Calculate the mass of the cluster Abell 1060 using the radial velocities of a sample of its members given in table 12.1. The cluster's effective radius is 100 arcmin and its distance is 68 Mpc.

Table 12.1 Radial velocities of members of the cluster Abell 1060 (velocities in units of 10^4 m s^{-1}).

451
362
369
246
380
357
251
264
456
375
274

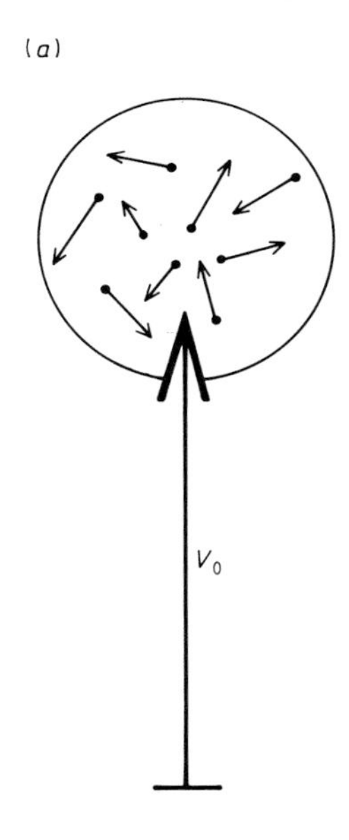

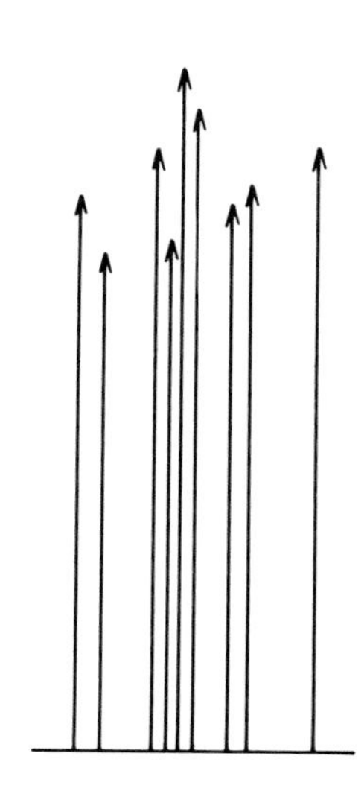

Figure 12.1 (*a*) Schematic representation of a cluster of galaxies receding with velocity V_0 in which individual members move at random in three dimensions. (*b*) Actual observed radial velocities of galaxies in the cluster. Each galaxy's velocity is made up of its own velocity in the line of sight and the velocity of recession of the cluster as a whole.

Suggestions. The velocity of recession of the cluster as a whole is the average of the velocities in the table. Subtract this average velocity from each of the individual velocities in turn to give each galaxy's own radial velocity within the cluster. Some of the residual velocities will be positive, some negative; but their sum, of course, will be zero. Square the residual velocities and take the mean of these numbers. This is V^2 in the formula.

Note that the formula for mass is in SI units. The value of R, the effective radius of the cluster, has therefore to be converted to metres. G in SI units is given in Appendix 1. The result of substitution in the formula will be the mass of the cluster in kilograms. To give the answer in solar units, divide by the mass of the Sun in kilograms.

THE AVERAGE MASS OF A GALAXY IN A CLUSTER

The average mass of a galaxy which follows from the total 'virial' mass of a cluster is a quantity of great importance. The discrepancy between individually derived masses of galaxies and virial masses is an indication of 'missing', or unobserved, mass in the Universe. There is evidence from galaxy rotation curves that galaxies contain additional material beyond their observed apparent limits. The shortfall is even more pronounced in clusters of galaxies and may indicate that there is unseen material in the spaces between the galaxies in clusters.

Exercise 5 (numerical). There are approximately 200 galaxies in the cluster Abell 1060 in the upper range of 4.5 magnitudes (that means in practice, galaxies down to approximately the luminosity of the Magellanic Clouds—see table 11.1). Estimate the average mass of a galaxy in the cluster in solar masses. Compare the result with the mass of our Galaxy and of other galaxies in the Local Group (§11).

SOLUTIONS AND DISCUSSION

1. Abell 1060 and Centaurus 2 are irregular. Pavo is regular; it has a giant elliptical galaxy in the middle, surrounded by many spirals.

2. Measurements to the nearest millimetre on the prints of the three largest elliptical galaxies and the two largest spirals in each cluster, converted to arcminutes from the scales, give ratios of sizes of 0.27, 0.18, 0.21, 0.31, 0.23. The mean is 0.24; the reciprocal is 4.2.

$\therefore$ distance $= 84 \pm 20$ Mpc.

This result, from measurements made by one individual, is not to be taken as exact; others will give different estimates of the dimensions of the galaxies. The distance agrees fairly well with the distance of 68 Mpc, which has been quoted for Abell 1060. Other astronomers favour a distance of 16 Mpc for the Virgo cluster which would reduce our figure to 67 Mpc. In precise work, the level of the isophotes is recorded photometrically to ensure that exactly the same level of brightness is being used in all cases.

The same exercise may be applied to the Centaurus II cluster; its distance is 75 Mpc.

3. The photograph was divided into 108 squares of arbitrary size. 76 squares at the edges represented regions outside the cluster.

Total counts of recognisable galaxies = 56
Number in 76 outer squares = 21, corresponding to 30 in 108 squares
$\therefore$ number in cluster = 26.
Estimated mean radius of cluster = 12 arcmin
Corresponding linear radius = 0.35 Mpc approximately.

The dimensions and number of galaxies are about the same as for the Local Group. However, most of the Local Group galaxies would be invisible at a distance of 100 Mpc; a more realistic number of comparable galaxies in the Local Group would be 6–8.

4. The mean velocity is 344×10^4 m s^{-1}. The mean of the squares of the residuals is 5.2×10^{11} (m s^{-1})^2

Radius of cluster $R = 6.1 \times 10^{22}$ m
$M = 3V^2R/G = 14 \times 10^{44}$ kg $= 7 \times 10^{14}$ solar masses (the published value).

5. If there are 200 galaxies, the average mass of a galaxy is 3.5×10^{12} solar masses. This is higher by a factor of 10 than the mass of the Andromeda or Milky Way galaxies. The result does not mean that the actual galaxies contain the extra mass; it may exist in some form in between the galaxies. The discrepancy has not been explained.

Note on distances. The velocities of recession of the galaxies increase with distance on account of the expansion of the Universe. The relation between velocity and distance follows Hubble's law: $V = Hr$ where V is velocity of recession in km s^{-1}, r is in Mpc, and H is Hubble's constant. The value of H is found from radial velocities of galaxies for which distances are known by independent means. The measurement of these distances is subject to many uncertainties, and it is for this reason that the value of Hubble's constant is so difficult to determine. Its value certainly lies between 50 and 100; the most generally accepted value is 60. Once a value of H is accepted, the procedure may be reversed, and the distance of a galaxy or a cluster of galaxies inferred from its radial velocity. It is the only method available for very faint distant objects.

In Exercise 4, the average of the radial velocities of member galaxies of the cluster Abell 1060 was found to be 344×10^4 m s^{-1} or 3440 km s^{-1}. The distance, by Hubble's law and with $H = 60$, is $3440/60 = 57$ Mpc, which differs from the value of 68 Mpc given in the exercise. This value was obtained by an independent method, the 'Tully–Fisher relation', which links luminosities of spiral galaxies with the widths of their 21 cm H I radio emission line. The width of this line is a measure of the Doppler displacement caused by the galaxy's rotation (observed directly in instances such as those discussed in § 11) which in turn is related to the mass of the galaxy.

Notice that all the results given in the various solutions involve distances. In Solution 3, the radius calculated for the cluster is proportional to the distance. In Solution 4 the mass is proportional to the radius, that is, to the distance. Distances of galaxies reckoned from their recession can range over a factor of two, depending on what value of H between 50 and 100 is preferred. It is interesting, however, that a variation of a factor of two still leaves a high average mass per galaxy in a cluster when calculated by the virial theorem.

Appendix 1 Astronomical Definitions, Data and Conversions

The numerical data in the tables are given to three significant figures, though two significant figures are generally sufficient for the exercises. In fact, rounded values (shown in brackets) of many of the basic quantities happen to be very simple numbers; these make order-of-magnitude calculations easy to work out and give answers compatible with the accuracy of most of the measurements.

Angles

degree = 60 arcminutes
= 3600 arcseconds
radian (rad) = 57.3 degrees
= 2.06×10^5 arcseconds (2×10^5)
π radians = 180°
steradian (sr) = 1 square radian
= 3.28×10^3 square degrees (3.3×10^3)
square degree = 3.05×10^{-4} steradian (3×10^{-4})
linear dimension of an object at distance r subtending an angle θ radians = $r\theta$
area of an extended object at distance r subtending a solid angle ω steradians = $r^2\omega$
volume of a cone of radius r with solid angle ω steradians at the apex = $\frac{1}{3}r^3\omega$

Time

year = 3.16×10^7 seconds (s) (3×10^7)
speed of light = 3.00×10^8 metres per second (m s^{-1}) (3×10^8)

Divisions and multiples of the metre (m)

nanometre (nm) = 10^{-9} m
micron (μ) = 10^{-6} m
millimetre (mm) = 10^{-3} m
kilometre (km) = 10^3 m

Astronomical distances

astronomical unit (AU) (Earth to Sun) = 1.50×10^{11} m (1.5×10^{11})
parsec (pc) = 2.06×10^5 AU (2×10^5)
= 3.09×10^{16} m (3×10^7)
kiloparsec (kpc) = 10^3 pc (3×10^{19} m)
megaparsec (Mpc) = 10^6 pc (3×10^{22} m)
light year = distance travelled by light in 1 year = 9.46×10^{15} m (9.5×10^{15} m)
1 parsec = 3.26 light years

Mass

Mass of the Sun = 1.99×10^{30} kilograms (kg) (2×10^{30})
Constant of gravitation = 6.67×10^{-11} SI units (6.7×10^{-11})

Appendix 2 Magnitudes, Colours and Luminosities

Magnitudes

Difference of 5 magnitudes = factor of 100 in brightness.
Magnitude difference between two stars, 1 and 2, of brightnesses b_1 and b_2 = $m_2 - m_1 = 2.5 \log (b_1/b_2)$.
$b_1/b_2 = 10^{0.4(m_2 - m_1)}$.
Absolute magnitude = the apparent magnitude which a star would have if seen at a distance of 10 pc.
Distance modulus $m - M = 5 \log (r/10) = 5 \log r - 5$, where m is apparent magnitude, M is absolute magnitude, and r is distance in parsecs.
Absolute magnitude of the Sun in the V waveband, $M_V = 4.83$.

Magnitude systems

The most widely used system in the optical range uses three wavebands, V (visual, in the yellow) centred on 0.55 microns, B (blue) centred on 0.44 microns and U (ultraviolet, meaning the near-ultraviolet which is received at the Earth) centred on 0.36 microns. Apparent magnitudes on the UBV system, as it is called, are labelled V, B and U. Absolute magnitudes on the same system are labelled M_V, M_B and M_U. Many of the UK Schmidt Telescope photographs, from the Southern Sky Survey, have been taken in a blue waveband labelled J which is close to the B band. Other wavebands, whenever used in the exercises, are described in the text.

Radiation fluxes

Apparent magnitudes may be translated to radiation fluxes on an absolute scale received from the star per unit area at the top of the Earth's atmosphere per unit of wavelength per second, from the standard values of the fluxes established for a star of apparent magnitude 0. The values corresponding to apparent magnitude exactly 0 are:

$$V = 0 \qquad f = 3.8 \times 10^{-11}\ \mathrm{W\ m^{-2}\ nm^{-1}}$$

$$B = 0 \qquad f = 6.6 \times 10^{-11}\ \mathrm{W\ m^{-2}\ nm^{-1}}$$

and similar quantities for other wavebands.

By definition of the magnitude system, the flux received from a star of magnitude m is $10^{-0.4m}$ times that from a star of magnitude 0, f_0. Expressed logarithmically, $\log f$ (for star of magnitude m) $= \log f_0 - 0.4m$.

Colours

A measure of the relative fluxes received at different wavelength regions in the spectrum is given by the magnitude differences at those wavelengths, $B - V$ and $U - B$, (and similar differences between other pairs of wavebands) which are called 'colour indices'. The difference $B - V$, often simply called the 'colour' of a star, is the most useful colour index for most purposes in optical astronomy.

The colour $B - V$ and all other indices of the standard star (which has magnitude 0 in all wavebands) are zero. This does not mean that the fluxes are equal at all wavelengths in the standard star; it simply means that the standard fixes the zero point of colour index. Stars for which $B-V$ is positive have relatively more red light in their spectrum than the standard star; those for which $B-V$ is negative have relatively more blue light; these stars are referred to as red or blue respectively.

Luminosities

The luminosity of a star is the total flux per second radiated by the star in all directions. The fluxes f given above refer to a width of one unit of wavelength in a particular waveband. The total flux reaching the top of the Earth's atmosphere from the standard star, summed up over all wavelengths including the spectral regions not actually observed, is

$$F=2.5\times10^{-8}\ \mathrm{W\,m^{-2}}.$$

The flux from a star of apparent magnitude 0 is by definition the same as that from a star of absolute magnitude 0 at a distance of 10 pc. If L is the star's luminosity and r is its distance, this flux spreads out over a sphere of radius r and area $4\pi r^2$, and therefore the flux per unit area arriving at the Earth is

$$F=L/4\pi r^2$$

The distance $r=10$ pc $=3\times10^{17}$ m. Substituting F and r gives

$$L=3.0\times10^{28}\ \mathrm{W}.$$

The luminosities of stars are usually given in units of the Sun's luminosity $L_{\odot}$,

$$L_{\odot}=3.8\times10^{26}\ \mathrm{W}.$$

Appendix 3 Sources and References

Photographs are copyright of the sources and are reproduced by permission.

1.1 and 1.2 Royal Greenwich Observatory
1.3 Instituto Astrofisica Canarias, La Laguna, Tenerife, Spain
5.1 Lick Observatory
5.2 Observatoire de Marseille
11.1 National Geographic Society—Palomar Sky Survey
All other photographs are from the UK Schmidt Telescope, provided by Photolabs, Royal Observatory Edinburgh.

REFERENCES

Data in tables 3.1 and 3.2 from D. K. Yeomans, *Halley Comet Handbook,* NASA 1981.
Description of the Halley's comet disconnection event in §3 from M. B. Neidner, Jr and K. Schwingenschuh, 1987, *Astron. Astrophys.* **187** 103.
Data in table 4.1 from G. Alcaino and W. Liller, 1964, *Astron. J.* **89** 814; in table 4.2 from K. P. Tritton, A. Savage and D. C. Morton, 1984, *Mon. Not. R. Astron. Soc.* **206** 843; in table 4.3 from C. Roslund, 1969, *Arkiv för Astronomi* **5** 249; in table 4.4 from C. W. Allen, 1973, *Astrophysical Quantities*, 3rd edition, London.
Data in table 6.1 from H. Johnson, 1958, *Astrophys. J.* **128** 33; in table 6.2 from C. W. Allen, 1973, *Astrophysical Quantities,* 3rd edition, London; data in figure 6.2 from G. Hagen, 1970, *Publ. David Dunlop Ob.* **4,** other data from F. D. A. Hartwick and J. E. Hesser, 1974, *Astrophys. J.* **192** 391.
Figure 8.4 from J. Mayo Greenberg, 1984, *Occ. Rep. R. Obs. Edinburgh* **12** 1 (with permission).
Figure 9.3 from R. W. Hanuschik, 1989, *Rev. Mod. Astron.* **2** 148 (with permission).
Figure 11.2 adapted from V. C. Rubin and W. K. Ford, 1970, *Astrophys. J.* **159** 379 (with permission).
Figure 11.3 from J. Hindman, 1967, *Australian J. Phys.* **20** 147 (with permission). Figure 11.4 from K. Newton and D. T. Emerson, 1977, *Mon. Not. R. Astron. Soc.* **181** 573 (with permission).
Data in table 12.1 and other data on cluster Abell 1060 (Hydra I) from O. G. Richter, J. Materne and W. K. Huchtmeier, 1982, *Astron. Astrophys.* **111** 193 and O. G. Richter and W. K. Huchtmeier, 1983, *Astron. Astrophys.* **125** 187.

Index